Living with the Georgia Shore

Living with the Shore
Series editors, Orrin H. Pilkey, Jr.,
and William J. Neal

The Beaches Are Moving: The Drowning
of America's Shoreline
New edition
Wallace Kaufman and Orrin H. Pilkey, Jr.

From Currituck to Calabash: Living with
North Carolina's Barrier Islands
Second edition
Orrin H. Pilkey, Jr., et al.

Living with the Alabama-Mississippi Shore
Wayne F. Canis et al.

Living with the California Coast
Gary Griggs and Lauret Savoy et al.

Living with the Chesapeake Bay and
Virginia's Ocean Shores
Larry G. Ward et al.

A Moveable Shore: The Fate of the
Connecticut Coast
Peter C. Patton and James M. Kent

Living with the Coast of Maine
Joseph T. Kelley et al.

Living with the East Florida Shore
Orrin H. Pilkey, Jr., et al.

Living with the Lake Erie Shore
Charles H. Carter et al.

Living with Long Island's South Shore
Larry McCormick et al.

Living with the Louisiana Shore
Joseph T. Kelley et al.

Living with the New Jersey Shore
Karl F. Nordstrom et al.

Living with the Shore of Puget Sound
and the Georgia Strait
Thomas A. Terich

Living with the South Carolina Shore
William J. Neal et al.

Living with the Texas Shore
Robert A. Morton et al.

Living with the West Florida Shore
Larry J. Doyle et al.

Living with the Georgia Shore

Tonya D. Clayton, Lewis A. Taylor, Jr., William J. Cleary, Paul E. Hosier,

Peter H. F. Graber, William J. Neal, and Orrin H. Pilkey, Sr.

Duke University Press   Durham and London   1992

Sponsored by the

National Audubon Society™

The Living with the Shore series

is funded by the Federal Emergency

Management Agency

The National Audubon Society
and Its Mission

In the late 1800s, forward-thinking people became concerned over the slaughter of plumed birds for the millinery trade. They gathered together in groups to protest, calling themselves Audubon societies after the famous painter and naturalist John James Audubon. In 1905, thirty-five state Audubon groups incorporated as the National Association of Audubon Societies for the Protection of Wild Birds and Animals, since shortened to National Audubon Society. Now, with more than half a million members, five hundred chapters, ten regional offices, a twenty-five million dollar budget, and a staff of two hundred seventy-three, the Audubon Society is a powerful force for conservation, research, education, and action.

The Society's headquarters are in New York City; the legislative branch works out of an office on Capitol Hill in Washington, D.C. Ecology camps, environmental education centers, research stations, and eighty sanctuaries are strategically located around the country. The Society publishes a prize-winning magazine, *Audubon,* an ornithological journal, *American Birds,* a newspaper of environmental issues and society activities, *Audubon Action,* and a newsletter as part of the youth education program, *Audubon Adventures.*

The Society's mission is expressed by the Audubon Cause: to conserve plants and animals and their habitats, to further the wise use of land and water, to promote rational energy strategies, to protect life from pollution, and to seek solutions to global environmental problems.

National Audubon Society
950 Third Avenue   New York, New York 10022

Living with the Shore Series

Publication of 16 volumes in the Living with the Shore series has been funded by the Federal Emergency Management Agency.

The Sapelo Island Research Foundation furnished special funding for the Georgia volume.

Publication has been greatly assisted by the following individuals and organizations: the American Conservation Association, an anonymous Texas foundation, the Charleston Natural History Society, the Office of Coastal Zone Management (NOAA), the Geraldine R. Dodge Foundation, the William H. Donner Foundation, Inc., the George Gund Foundation, the Mobil Oil Corporation, Elizabeth O'Connor, the Sea Grant programs in New Jersey, North Carolina, Florida, Mississippi/Alabama, and New York, The Fund for New Jersey, M. Harvey Weil, Patrick H. Welder, Jr., and the Water Resources Institute of Grand Valley State University. The Living with the Shore series is a product of the Duke University Program for the Study of Developed Shorelines, which was initially funded by the Donner Foundation.

©1992 Duke University Press

All rights reserved

Printed in the United States of America

on acid-free paper

Library of Congress Cataloging-in-Publication Data

appear on the last printed page of this book.

Contents

Figures and Tables ix

Preface xi

Prologue xiii

Barrier Islands xiii
  The Origins of Barrier Islands: Georgia's Two Generations xiii
  Shifting Shores xiii

1 Introduction: A Coastal Perspective 1

Threats at the Shore 3
Where to from Here? 5

2 The Guale Coast: A Human Perspective 7

"The Fairest, Fruitfullest, and Pleasantest . . ." 7
Island Management: The Perspectives of the Present 8

3 The Guale Coast: A Natural Perspective 13

The Coastal Setting 13
The Ups and Downs of Sea Level 14
The Greenhouse Effect and Sea-Level Rise 18
Barrier Island Processes: An Integrated System 19
  Beaches: Their Dynamic Equilibrium 19
  Origin of Sands of the Shoreline 19
  Moving Sand: Wind, Waves, and Currents 20
    Winds and Waves 20
    Tides 21
    Erosion—Why and Where Does the Sand Go? 23
  How to Recognize an Eroding Shoreline 24
  Accretion—The Beach Widens 25
  Island Migration 27
  Storms and Beaches—The Sand-Sharing System in Action 27
Coastal Storms—Hazards, Headlines, and Heartbreaks 28
  Nor'easters 28
  Hurricanes: Killers at the Coast 30
    Storm Surge 31
    Hurricane Probability 32
    Hurricanes in Georgia 33
    Good News: The Georgia Plan 35
    Stormhounds: A Word of Caution 36
Of *Smilax*, Sabal, and *Spartina*: Patterns of Barrier Island Vegetation 36
  A Walk Across an Island 36
Coastal Conflicts: The Collision of Human Plans and Natural Processes 38
  Learning to Live with an Island 39

4 Human Nature and Nature's Shores 41

Shoreline Engineering: Trying to Stabilize the Unstable 41
  Seawalls and Bulkheads 42
  Riprap and Revetments 44
  Groins 45
  Jetties 47
  Breakwaters 47
  Beach Replenishment 48
  Whose Cost? Whose Benefit? 51
  A Modest Proposal 51
  Questions to Ask If Shoreline Engineering Is Proposed 52
Truths of the Shoreline 54
  Some Guiding Principles 55

5 Selecting a Site on Georgia's Barrier Islands 57

Individual Island Analysis 57
Selecting "Your" Island 58
  Physical Factors 58
    On an Island 58
    On the Beachfront 60
    On a Riverbank 62
  Biological Factors 63
  Political and Infrastructure Factors 64
Site Safety Checklist 66
The Georgia Islands 69
  The Rating System 69

Tybee Island 70
  North Tybee Island 72
  Central Tybee Island 73
  South Tybee Island 74
  History of Erosion Control Efforts 76
  Evacuation 79
  Services 80
  Land Use 80
Little Tybee and Williamson Islands 80
  Little Tybee 81
  Williamson Island 81
Wassaw Island 82
Ossabaw Island 83
St. Catherines Island 84
Blackbeard Island 86
Sapelo Island 87
Wolf Island 88
Little St. Simons Island 88
  Elevation 88
  History of MHW Shoreline Changes 89
  Inlet Migration 89
  Storm Overwash 89
  Storms 89
  Evacuation 89
  Land Use 89
Sea Island 90
  Elevation 92
  History of MHW Shoreline Changes 92
  Hampton River Shore 93
  Northeast Point to North End of Spit 93
  Spit Shoreline 94
  History of Erosion Control Efforts 94
  Inlet Migration 95
  Storm Overwash 95
  Storms and Hurricanes 95
  Evacuation 95
  Services 95
  Land Use 96
St. Simons Island 96
  Elevation 97
  History of MHW Changes 98
  Goulds Inlet Shore 98
  East Beach Area 98
  Southeastern and St. Simons Sound Shores 98
  History of Erosion Control Efforts 98
  Inlet Migration 100
  Storm Overwash 100
  Storms 100
  Evacuation 101
  Services 101
  Land Use 101
Jekyll Island 101
  Elevation 104
  History of MHW Changes 104
  St. Simons Sound Shore 104
  Oceanfront Strand 104
  South End 105
  History of Erosion Control Efforts 105
  Inlet Migration 105
  Storm Overwash 105
  Storms 105
  Evacuation 106
  Services 106
  Land Use 106
Little Cumberland Island 107
  Elevation 107
  History of MHW Changes 107
  Inlet Migration 107
  Evacuation 107
  Land Use 107
Cumberland Island 107
The Shore Challenge 110

6 Coastal Land-use Planning and Regulation 111

Coastal Marshlands Protection Act of 1970 111
  When Do I Need a Permit? 111
  What Must Be in the Application? 111
  How Is the Application Processed? 112
  Guidelines for Permit Evaluations 112
  Decisions on Permit Applications 113
  Appeal and Enforcement Procedures 113
  Term and Transfer of Permit 113
Shore Assistance Act of 1979 114
  When Do I Need a Permit? 114
  What Must Be in the Application? 114
  How Is the Application Processed? 115

Guidelines for Permit Evaluations 115
Shore Assistance Committee Meetings 115
Appeal and Enforcement Procedures 115
National Flood Insurance Program (NFIP) 116
Coastal Barrier Resources Act of 1982 and Coastal Barrier Improvement Act of 1990 117
Other Federal Programs 118
Summary 118
Recent Additions 118

7 Building or Buying a House Near the Shore 121

Coastal Realty Versus Coastal Reality 121
The Structure: Concept of Balanced Risk 121
Coastal Forces: Design Requirements 122
    Wind 122
    Storm Surge and Flooding 123
    Waves 124
    Battering by Debris 124
    Barometric Pressure Changes 124
House Selection 125
Keeping Dry: Pole or Stilt Houses 125
An Existing House: What to Look For, What to Improve 128
    Geographic Location 129
    How Well Is the House Built? 129
        Ground Anchorage and Foundation 129
        Roof 131

        Framing and Design 132
        Brick, Concrete, and Masonry Reinforcement 134
        Glass Surfaces 134
        In-House Shelter 135
    Summary: Protective Steps for Strengthening Buildings 135
Mobile Homes: Limiting Their Mobility 136
High-Rise Buildings: The Urban Shore 137
Modular Unit Construction: Prefabricating the Urban Shore 139
Living with Nature: Prudence Pays 140

Appendixes

A Hurricane Checklist 141

B Recent Hurricanes in Georgia 144

C A Guide to Federal, State, and Local Agencies and Organizations Involved in Coastal Development 146

D Useful References 159

E Selected Field Stops on Tybee Island 179

Index 187

Figures and Tables

Figures

1.1 (a) Natural Georgia coast, (b) Developed Georgia coast 1
1.2 Index map of the Georgia coast 2
1.3 Undeveloped Sapelo Island 4
1.4 Cluttered New Jersey shore 5
1.5 Sea Island "Private Property" placard 5
2.1 Tybee Island seawall 9
2.2 Riprap seawall on St. Simons Island 10
2.3 Sea Island seawall 10
2.4 Jekyll Island seawall 11
3.1 Physiographic provinces of Georgia 13
3.2 Bathymetry of the continental shelf adjacent to the Georgia coast 14
3.3 (a) Monthly sea level graph, (b) Annual sea level graph 14
3.4 Fluctuations in sea level during the past 40,000 years 15
3.5 Relationship between sea-level rise and shoreline movement 15
3.6 Georgia coast showing Holocene and Pleistocene islands 16
3.7 Georgia's lower coastal plain in cross section, showing former barrier islands 17
3.8 One scenario for the formation of barrier islands 17
3.9 Landward movement of Holocene barrier islands with sea level rise 18
3.10 The dynamic equilibrium of the beach 19
3.11 Creation of longshore currents 21
3.12 Georgia's eight major island units 21
3.13 Spring tides and neap tides 22
3.14 Typical tidal delta on the Georgia coast 22
3.15 Ebb tidal delta at Tybee Creek inlet 23
3.16 Evidence of shoreline recession 25
3.17 Wassaw Island shoreline recession 26
3.18 Historic migration of Goulds Inlet 26
3.19 Barrier island migration at Assateague Island, Virginia 27
3.20 Beach flattening in response to a storm 28
3.21 Sea Island seawall after 1981 storm 29
3.22 Washover activity on the Georgia coast 30
3.23 Brunswick, Georgia, after the 1898 hurricane 33
3.24 Historic hurricane tracks across Georgia's coast 35
3.25 Typical Georgia island vegetation zonation 37
4.1 Hazards created by shoreline engineering 41
4.2 Natural beaches on Little Tybee Island 41
4.3 Tybee Island Dune Restoration Project 42
4.4 South end of Tybee Island 42
4.5 Sea Island sloping seawall 43
4.6 Erosion control fortifications on St. Simons Island 44
4.7 Seawall with riprap at toe, Sea Island 45
4.8 Effect of groins 46
4.9 Terminal groin on Tybee Island 46
4.10 Overhead view of a breakwater 47
4.11 Beach replenishment 48
4.12 Tybee Island beach replenishment 49
5.1 Savannah River industry 60
5.2 (a) Eroding dune, (b) Natural dune field on Jekyll Island, (c) Artificial dunes on Tybee Island 61
5.3 Overwash fan on north end of Jekyll Island 62
5.4 Eroding riverbank on the Cumberland River 63
5.5 Finger canal on Tybee Island 63
5.6 Mature dune vegetation and stabilized dune system on Jekyll Island 64
5.7 Causeway to Tybee Island 65
5.8 Site analysis map of Tybee Island 71
5.9 Historic net erosion and accretion on Tybee and Little Tybee Islands 72
5.10 Aerial view of north end of Tybee Island 73
5.11 Butler Avenue/Highway 80 vicinity of Tybee Island: (a) Replenished section with no high tide beach, (b) Flooding behind the seawall 74
5.12 Central Tybee Island 74
5.13 Southern tip of Tybee Island with seawall and groins 75
5.14 (a) Tybee Island south parking lot, (b) The same parking lot, after the 1987 beach replenishment 75
5.15 South end of Tybee Island 77
5.16 South end of Tybee Island before the 1987 beach replenishment 78
5.17 South end of Tybee Island several months after the 1987 replenishment 78
5.18 Historic net erosion and accretion on Wassaw Island 82

5.19 Fort Morgan, Wassaw Island 82
5.20 Historic net erosion and accretion on Ossabaw Island 83
5.21 Site analysis map of St. Catherines Island 84
5.22 Historic net erosion and accretion on St. Catherines Island 86
5.23 St. Catherines "sea bluff" 86
5.24 Historic net erosion and accretion on Blackbeard and Sapelo Islands 87
5.25 Historic migration of Cabretta Inlet
5.26 Cabretta Island beach 88
5.27 Historic net erosion and accretion on Wolf Island 88
5.28 Historic net erosion and accretion on Little St. Simons Island 89
5.29 Site analysis map of Little St. Simons, Sea, and St. Simons Islands 90
5.30 Historic net erosion and accretion on Sea Island and St. Simons Island, (a) 1857/60 to 1974, (b) 1924 to 1974 92
5.31 Sea Island sloping concrete wall 93
5.32 Sea Island revetment 94
5.33 Goulds Inlet area of St. Simons Island 98
5.34 Aerial view of East Beach area of St. Simons Island 99
5.35 East Beach 99
5.36 (a) St. Simons Island shore in the 1930s, (b) The same shore a half century later 100
5.37 Site analysis map of Jekyll Island 102
5.38 Historic net erosion and accretion on Jekyll Island 104
5.39 Jekyll Island "beach" 105
5.40 Site analysis map of Little Cumberland and Cumberland Islands 108
5.41 Historic net erosion and accretion on Little Cumberland and Cumberland Islands 109
7.1 Forces to be reckoned with at the shore 123
7.2 Modes of failure and how to deal with them 124
7.3 Shallow and deep supports for poles and posts 125
7.4 Framing system for an elevated house 127
7.5 Tying floors to poles 128
7.6 Additional connection of wood-frame building to foundation 129
7.7 Foundation anchorage 130
7.8 Stud-to-floor, plate-to-floor framing methods 130
7.9 Roof-to-wall connectors 130
7.10 Where to strengthen a house 131
7.11 Bracing walls at right angles to loaded surface 133
7.12 Some rules in selecting or designing a house 133
7.13 Reinforced tie beam (bond beam) for concrete block walls 134
7.14 Bathroom shelter module 135
7.15 Tiedowns for mobile homes 138
7.16 Hardware for mobile home tiedowns 139

Tables

3.1 Some Factors Affecting Sea Level 15
3.2 Georgia Barrier Island Sequences 16
3.3 Monthly Frequency of Tropical Cyclones in Georgia, 1898 to 1968 31
3.4 Saffir/Simpson Hurricane Scale 32
3.5 Significant Georgia Hurricanes 34
3.6 Plant Zonation on a Georgia Barrier Island 38
4.1 Methods Used to Reduce Erosion and Erosion Loss 53
5.1 History of Erosion Control Efforts at Tybee Island 66
5.2 History of Erosion Control Efforts at St. Simons Island 67
5.3 History of Erosion Control Efforts at Sea Island 68
5.4 History of Erosion Control Efforts at Jekyll Island 68
5.5 Summary of Conditions Considered in the Assignment of Risk Classifications 69
5.6 Tybee Island Shoreline Hazard-Risk Evaluation 70
5.7 St. Catherines Island Shoreline Hazard-Risk Evaluation 85
5.8 Little St. Simons Island Shoreline Hazard-Risk Evaluation 91
5.9 Sea Island Shoreline Hazard-Risk Evaluation 91
5.10 St. Simons Island Shoreline Hazard-Risk Evaluation 97
5.11 Jekyll Island Shoreline Hazard-Risk Evaluation 103
5.12 Little Cumberland Island Shoreline Hazard-Risk Evaluation 109
7.1 Tiedown Anchorage Requirements 137

Preface

We began our teaching careers in Georgia at the University of Georgia and Georgia Southern College (now Georgia Southern University), respectively. Research at the University of Georgia's Marine Institute on Sapelo Island and the Skidaway Institute of Oceanography took us to the barrier islands, estuaries, and marshes of this golden coast. We vacationed with our families on Jekyll Island and at Savannah Beach. Our children's first lessons in history were brought to life by visits to St. Simons Island, Fort Pulaski, or the reenactment of Oglethorpe's sail up the Savannah River. Their first experiences with nature were observations on the barrier islands and associated forests, marshes, creeks, and beaches. We hold the Georgia coast in special esteem, as did those who came before us. Our hope is that people in the future will have the same opportunities for discovery and learning.

This book's goal is to share some of what is known about the Georgia coast: to provide a perspective that will encourage prudence in the management and use of this dynamic area and to suggest that we make the greatest effort to live in harmony with the coast. This book is not meant to discourage development; we hope that it encourages proper, limited development. Although certain natural areas warrant protection from development, preservationism on all of the coast is an unrealistic philosophy, especially when a development pattern is already established in some places. Unrestricted development, however, endangers coastal residents and resources. The public should become aware of and concerned about our important coastal resources in order to conserve them.

The book, therefore, is primarily directed to the coastal dweller and the developer, to people who have built or bought on the shoreline or anywhere on a barrier island, and to those who plan to do so. It also should be valuable to real estate agents, bank loan officers, contractors, and investors who deal with the coast. Our goal is to increase the reader's awareness of how barrier islands may retreat, what kinds of hazards are faced by coastal dwellers and property owners, and how to reduce the effects of those hazards if you already live or work in such a zone.

The information presented here summarizes the work of many investigators, too numerous to mention, to whom we owe sincere thanks. Their respective agencies include Georgia Sea Grant, the Georgia Geological Survey, and the Skidaway Institute of Oceanography.

This book is one of some 20 projected volumes in the Living with the Shore series. Eventually, there will be a book for each coastal state as well as for Lake Erie and Lake Michigan.

As an umbrella book to the series, Duke University Press has reprinted with an updated appendix the classic *The Beaches Are Moving: The Drowning of America's Shoreline* by Wallace Kaufman and Orrin H. Pilkey, Jr. *The Beaches Are Moving* covers the basic issues dealt with specifically in the state-by-state books.

The series editors have published with Van Nostrand Reinhold Company a construction guide, *Coastal Design: A Guide for Builders, Planners, and Homeowners* (1983), giving detailed coastal construction principles. The prudent coastal dweller should read both *Coastal Design* and the individual state volume.

The overall coastal book project is an outgrowth of initial support from the National Oceanic and Atmospheric Administration through the Office of Coastal Zone Management. The project was administered through the North Carolina Sea Grant Program. Most recently we have been generously supported by the Federal Emergency Management Agency (FEMA). FEMA's support has enabled us to expand the project into a nationwide series including Lake Erie, Lake Michigan, and Puerto Rico. Without this support the series would have long since ground to a halt. The technical conclusions presented herein are those of the authors and do not necessarily represent those of the supporting agencies.

The Sapelo Island Research Foundation specifically contributed to producing the Georgia volume. For this support we are especially grateful.

In addition, a Gulf Coast Oceanographic Charitable Trust Fellowship and a William and Elsie Knight Fellowship, granted to the senior author through the University of South Florida Department of Marine Science, provided for critical and incidental expenses that kept production efforts going in the final days of preparing the manuscript.

We owe a debt of gratitude to many individuals for support, ideas, encouragement, and information. Doris Schroeder has helped us in many ways as Jill-of-all-trades over a span of more than a decade and a dozen books. Barbara Gruver, Terri Rust, and Lynne Claflin drafted figures for this volume. Amber Taylor also drafted figures and handled many last minute details. Debbie Gooch patiently typed and retyped many drafts of the text. The original idea for our first coastal book (*How to Live with an Island*, 1972) was that of Pete Chenery, then director of the North Carolina Science and Technology Research Center. Richard Foster of the Federal Coastal Zone Management Agency supported the book project at a critical juncture. Because of his lifelong commitment to land conservation, Richard Pough of the Natural Area Council has been a mainstay in our fund-raising efforts. Myrna Jackson of the Duke Development Office and the President's Associates of Duke University have been most helpful in our search for support.

Jane Bullock, Gary Johnson, Richard Krimm, Fred Sharrocks, and Gene Zeizel of the Federal Emergency Management Agency have worked hard to help us chart a course through the shifting channels of the federal government. Bob Weiss and Eric Rosenberg, Peter Gibson, Dennis Carroll, Jim Collins, Jet Battley, and many others opened doors, furnished maps and charts, and in many other ways helped us through the Washington maze.

In Georgia we were assisted by the staffs of the State Coastal Resources Division of the Department of Natural Resources and local historical societies and museums, as well as the coastal chapter of the Audubon Society. Kay Benson of St. Simons Island was especially helpful and hospitable, acting as local hostess and tour guide. Many other coastal residents, too numerous to name, also lent a hand; to all we are grateful.

Orrin H. Pilkey, Jr.
and William J. Neal
Series Editors

Prologue

Perhaps you're considering the island life as the perfect answer to those big city blues, or maybe as part of your retirement plans. Perhaps you already own a house or property as an island retreat. Or perhaps you're just a visitor to the coast—content to leave only footprints and take only memories. Regardless of the nature of your relationship to the islands, this book has been designed with you in mind.

Living "with" a coast is much different from simply living "on" one. The former requires forethought, planning, and wise choice of property or a house site. The latter, which many coastal residents follow, doesn't allow for the potential hazards of coastal life or for the coast's rapidly changing nature. The safety and longevity of your island home, even your life, could directly depend on the choices you make about the proper location and construction of your home.

Living with the Georgia Shore is a reference and guide for residents and for visitors to the Georgia coast. Coastal perspectives—historical, natural, engineering, and legal—provide a basis for understanding the workings of coastal systems. These principles are then incorporated into a set of guidelines for evaluating coastal hazards on Georgia islands. The resulting portrait illustrates some appropriate planning and selection strategies for developing an island life-style that blends with coastal processes and systems.

Barrier Islands

Seaward of almost every gently sloping coastal plain in the world lie narrow ribbons of sand called *barrier islands*. Most of the U.S. Atlantic and Gulf shores, for example, are rimmed by these sandy islands—about 300 in all. (The number is constantly changing, as islands appear, disappear, and reappear, and inlets open, close, and move about.)

The Origin of Barrier Islands: Georgia's Two Generations

Modes of barrier island formation vary with local differences in geological processes and coastal landforms. In general, two key ingredients are required for the creation of barrier islands: (1) a gently sloping sandy coastal plain and (2) a rising sea level. The Georgia coast has both, as does much of the U.S. Atlantic and Gulf coastline. However, the coasts of Georgia, southern South Carolina, and a small bit of northern Florida also have an island system with two generations of barriers at the sea's edge.

The older barrier islands are called *Pleistocene* in reference to the geologic period in which they were formed. The "new" barrier islands are referred to as recent (or *Holocene*).

In general, the recent islands represent a relatively narrow strip of land on the ocean side of the islands. In other words, most of the bulk and area of Georgia's islands represent ancient islands.

Shifting Shores

The current configuration of the Georgia barrier islands is only the most recent in a series of changing shorelines. From the fall line to far out on the continental shelf are locations of former shorelines. Whereas the oldest of the relict shorelines we can identify today dates back to the Miocene Epoch (12 to 25 million years ago), several shorelines located by scientists were created more recently, during the Pleistocene Epoch (1.8 million years to 10,000 years ago).

Fifty to 60 miles west of the present Atlantic shore lies the oldest of these relict Pleistocene shorelines at 110 feet above mean sea level (MSL). Six such parallel shoreline strands decorate the coastal plain. In cross section, the region resembles a series of steps, or terraces, descending to the ocean.

These relict shorelines were created during an episode of falling sea level. Each is located where an ancient seaward-marching shoreline was temporarily "stabilized" as the rate of sea-level fall slowed or stopped. If sea level subsequently rose slightly, barrier islands might have been formed; river-borne sediments from uplands could fill back-barrier lagoons and create expansive salt marshes. Instead, when the sea level began to fall again, receding from the land, a terrace of former barrier islands and salt marshes was left high and dry on the newly exposed coastal plain. Scientists today recognize these terraces as markers of former shorelines. Thus, the edge of the current mainland is marked by remnants of a Pleistocene shoreline.

The one factor that is a constant for a coast also applies to the material in this book. That factor is *change*. Change, the dynamic nature of all systems, will always characterize coastal processes, island shapes and locations, regulations and laws. Some of the hazard evaluations and coastal regulations in this book, therefore, may become inapplicable after its publication. Although such changes will necessitate minor modifications in the book's use, we have provided a framework that allows for such flexibility.

Just as coastal processes interact to sculpt the shape of an ocean beach, several individuals combined their efforts to mold this book. The text results from several years of toil and turmoil, and several persons guided and contributed to the process. The book is part of the Living with the Shore series produced by Duke University Press. Orrin Pilkey, Jr., and William Neal, the instigators and senior editors for the series, are responsible for the final editing. The one person who ensured that this book was produced is Tonya Clayton, the coordinator and a major author. We thank Tonya for her gentle coercion and coaxing of each of the contributors to complete long, often overdue, assignments. The final publication strongly reflects her influence and contributions.

In addition, the following authors made integral contributions. Bill Cleary, a marine geologist, and Paul Hosier, a plant ecologist, both from the University of North Carolina at Wilmington, prepared the initial framework and working draft. They served as critical consultants during the text's maturation. Peter Graber, a lawyer who has published extensively on coastal law and regulation, produced the chapter on coastal land-use planning and regulation. Lewis Taylor, a former science educator with the University of Georgia's Marine Extension Service, contributed material for the introduction, the sections on natural and historical perspectives of the coast, and the site selection and hazard risk assessment sections, as well as the field guide to Tybee Island. Orrin Pilkey, Sr., a retired civil engineer, wrote the chapter on coastal construction techniques. The book's strength lies in the combined expertise of each of these contributors.

Sincere thanks are also owed to several individuals for their support and assistance during the development of this project. Jim Howard, a geologist who has influenced our present understanding of the geology of the Georgia coast, provided much wisdom and encouragement. Meredyth McIlwaine also has remained a source of encouragement and assistance with fieldwork.

In addition to these two key supporters, we have received constructive criticism and assistance from many others. Hans Neuhauser, director of the Georgia Conservancy, and Vernon J. Henry, head of the Geology Department of Georgia State University, served as invaluable consultants and reviewers. Several other individuals and agencies provided assistance, consultations, and critical reviews. Will Hon, Carol Johnson, and Taylor Schoettle, teaching staff for the Marine Extension Service of the University of Georgia, insightfully evaluated portions of the manuscript. Members of the Marsh and Beach section of the Coastal Resources Division of the Georgia Department of Natural Resources, many individuals with the Savannah District of the U.S. Army Corps of Engineers, the administrative staff for the City of Tybee, the helicopter pilots of the U.S. Coast Guard Search and Rescue Division in Savannah, and Chap Norton, pilot for the Chatham County Mosquito Control, aided in these efforts. And, of course, Tonya Clayton, Orrin Pilkey, and William Neal provided the primary and final impetus for completion of the work.

Living with the Georgia Shore

1 Introduction: A Coastal Perspective

"Coast." The word elicits a myriad of memories and feelings among those who have been touched by the power of the ocean's kiss on the land. A beach, or an island, is usually the first image conjured up by thoughts of coastal areas. For some, the coast is thought of in terms of its natural components—deserted, windswept beaches; liana-laden island forests; broad expanses of water, marsh, and sky. For many others the coast is the place for sand castles and surfing, boardwalks and beach houses, rest and recreation (fig. 1.1). In either case, the fundamental attractions remain the sea, sand, and sun.

The barrier islands of the Georgia coast (fig. 1.2) are renowned for their unique beauty. A large tidal range, low wave energy, and fine sands allow for broad expanses of firm beaches—a paradise for beach aficionados. And the available beach experiences range from the hustle and bustle of Tybee Island's beachfront boardwalk to the solitude of Cumberland Island's wilderness. Through the kind fortune of events and locale, more than of planning and wisdom, the Georgia coast provides a barrier island chain with a rich variety of island experiences.

The natural features and processes of Georgia's islands, as well as those of southern South Carolina and northern Florida, differ substantially from the rest of the barrier islands of the U.S. shoreline. Unlike the long and narrow islands of wave-dominated coasts such as those found in North Carolina or Texas, islands along Georgia's shores are short and wide. Extensive maritime forests blanket the Georgia barriers—a function of the greater width and relative "stability" of the islands (fig. 1.3). Change in island position and shape, however, is still the constant that dominates the coastline's evolution.

These same island dynamics directly affect the long-term stability of island residences and the safety of their inhabitants. Episodic events such as hurricanes and winter storms pose the most serious hazards to living on the islands. The long-term effects of shoreline recession (erosion), coupled with the prospect of an increasing rate of sea-level rise, translate into major problems and increased costs for island residents, visitors, and Georgia taxpayers.

At present, most of the Georgia islands, unlike those of neighboring states, are uninhabited places of coastal wilderness (fig. 1.3). Nine of the 13 barrier islands exist in this condition, a fact that can be credited to unusual historical ownership patterns, low population pressures, and the lack of automobile access. The "developed" barriers of the coast, all of which can be reached by car, are those on which this book focuses: Tybee, Sea, St. Simons, and Jekyll Islands.

1.1 (a) Natural Georgia coast. Miles of uninterrupted wide beach backed by dunes and maritime forest. Jekyll Island, early 1980s. Photo by W. Cleary. (b) Developed Georgia coast. Fun in the sun on Tybee Island, 1981. Photo by O. Pilkey.

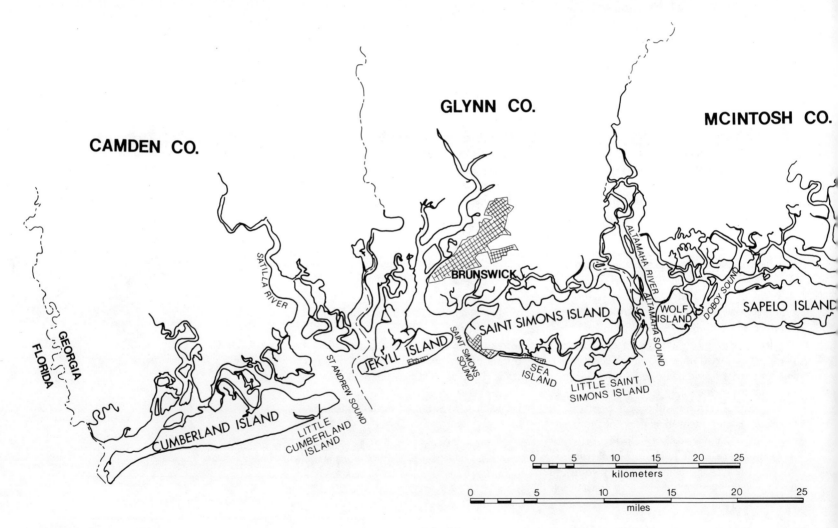

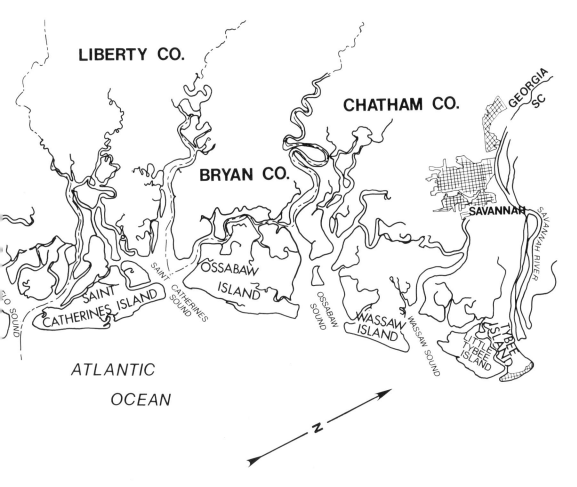

1.2 Index map of the Georgia coast.

Threats at the Shore

A 1976 study of land-use patterns along the U.S. coast found that approximately 5 percent of the acreage on Georgia's barrier islands was urbanized and that the growth rate in these urbanized areas during the previous two decades was 63 percent. Most of this development was on Tybee, Sea, St. Simons, and Jekyll Islands. Yet compared to other coastal states, the proportion of Georgia's islands that was considered "developed" was relatively low, as was the rate of urbanization. For instance, North Carolina's islands had 14 percent of their area developed, an increase of 269 percent during the previous 20 years! And New Jersey's islands, while experiencing an increase of only 28 percent in urbanized acreage, were considered to be developed on more than 47 percent of their land. Thanks to the protected status of most of the barriers, the majority of Georgia's islands should be spared many problems typically associated with rapid and extensive unplanned development.

Nevertheless, the situation on the developed Georgia islands may continue to worsen as people attempt to "hold the line" against the sea. The relatively small degree of urbanization on the Georgia coast has not spared these developed islands from the problems that result when inflexible human development conflicts with dynamic natural systems. Although the

Introduction: A Coastal Perspective 3

1.3 Sapelo Island, an undeveloped island. View is of south end of Nanny Goat Beach, 1987. Such areas of coastal wildness are becoming a rarity in the contiguous United States. Photo by L. Taylor.

Georgia barriers in general do not have the severe problems of the New Jersey coast (fig. 1.4), the developed islands possess stretches of beach on which similar situations exist. Approximately 55 percent of the 19 miles of beach shoreline on Tybee, Sea, St. Simons, and Jekyll Islands has been armored with either concrete seawalls or revetments of granite boulders in response to the erosion of island uplands. On some sections of these islands the once pristine beaches now offer scenes akin to those on the New Jersey shore, where little or no beach area is available at high tide for recreational use. Continued acceleration of beach retreat (shoreline recession) threatens to worsen these problems. Already, expensive renourishment projects are needed to maintain artificial recreational beaches on Tybee Island and Sea Island, and more projects are being initiated and considered for the other islands.

Beach erosion, however, is only one of many processes that may create problems on the Georgia barriers. The quality of our life and environment, both wild and developed, depends on the ways we handle the following potential hazards:

Hurricanes and winter storms pose the greatest dangers to life and property. The Georgia coast has not experienced a major hurricane since 1898! The question we face, however, is only *when,* not *if,* another major hurricane will strike. Hazardous building sites and poor construction have created potentially unsafe situations on all of the developed Georgia islands. Effective evacuation before the storm hits still remains the single most important element in hurricane safety. Overpopulation of an island must be avoided to allow enough time for all residents to evacuate.

Sea-level rise is a driving force in the continued recession of the island shorelines. With the rate of sea-level rise expected to accelerate in conjunction with the "greenhouse effect," the severity and extent of shoreline problems will grow worse. A higher probability of hurricanes, more frequent coastal flooding, and greater vulnerability to coastal storms and subsequent shoreline erosion are likely to have a strong impact on Georgia barriers within the next few decades.

Pollution cannot be overlooked. Overdeveloping an island endangers the drinking water supply and taxes the natural cleansing abilities of the islands, adjacent marshes, and sounds. Improper waste disposal and even drainage of rainwater from roads and parking lots threaten the health of coastal citizens and

1.4 Cluttered New Jersey shore. Photo by O. Pilkey.

destroy the natural habitat that supports local marine fisheries. Shellfish beds are being closed at a disturbing rate throughout the southeastern United States.

Public access to beaches is another concern for Georgia residents. Private development inevitably reduces beach access to the general public. Often, adjacent upland property owners are the only ones with easy access to even public beach areas (fig. 1.5). (In Georgia, the area below the mean high water line is public property.) Adding insult to injury, publicly funded projects often end up paying the costs for storm damage repairs and beach nourishment.

Where to from Here?

The future of the Georgia barrier islands, especially the developed islands of Tybee, Sea, St. Simons, and Jekyll, depends on the ability of the state, counties, and local communities to evaluate the needs of and potential challenges to the islands. Differences in local laws and their implementation strongly affect the quality of life on the individual Georgia barrier islands. For that reason, state, regional, and local island management strategies, zoning laws, and land-use regulations must be strong, coordinated, and enforced to effectively guide future growth and cope with beach erosion, coastal storms, construction and post-storm reconstruction, beach access, and sea-level rise.

These problems confront coastal dwellers nearly everywhere. Unfortunately, the same mistakes in attempting to solve or reduce such problems are often repeated. Today's coastal residents, developers, and managers have the opportunity to learn from the mistakes of others, to avoid many coastal hazards, and to chart a safer course for coastal development.

During the past several decades a great deal has been learned about nearshore processes, coastal engineering, and the consequences of

1.5 This placard is placed on the windshields of cars parked on the streets of Sea Island. Restricted public access usually accompanies development of coastal areas.

NOTICE

DUNES AND BEACH ARE PRIVATE PROPERTY

Do Not Trespass
For Use Only by Sea Island Cottagers and
Cloister Hotel Guests.

--- Sea Island Co.

Introduction: A Coastal Perspective

developing coastal areas for private homes, condominiums, harbor facilities, and industry. With this information in hand, wise development is possible. Risks and environmental damage can be minimized. We can conserve the amenities that originally brought us to our paradise.

Our intent in these pages is to help residents, developers, and managers avoid or minimize the potential pitfalls of living with the shore. Chapter 2 provides a perspective on the human history of the islands—how their development evolved. Chapter 3 presents a natural history of island origin and change—how natural processes continue to shape the Georgia coast. Chapter 4 outlines the action/reaction when human activity and natural processes clash—how the knowledge of fundamentals can lead to recognizing "truths of the shoreline" (with respect to engineering) and guide one's decision-making. Chapter 5, which includes notes on historical shoreline responses to natural processes and man-made structures, provides a basis for selecting building sites with an eye to safety—how to live with the Georgia coast with minimal threat to property as well as human health. Chapter 6 capsulizes existing federal and state regulatory programs—where to begin in the permit application process. Chapter 7 concludes the text with advice on constructing or improving dwellings within the coastal zone—what to look for in a cottage, condominium, or mobile home. Finally, the appendixes provide safety information on hurricanes and storms (appendixes A and B), sources of information pertinent to living near the shore (appendix C), and useful references (appendix D). A field guide to Tybee Island (appendix E) provides a tour to see firsthand what this book is about. We also recommend H. E. Taylor Schoettle's *A Field Guide to Jekyll Island* (for details, see appendix D).

2 The Guale Coast: A Human Perspective

"The Fairest, Fruitfullest, and Pleasantest..."

In 1562 Frenchman Jean Ribaut described the Georgia coast as "the fairest, fruitfullest, and pleasantest" region he had ever seen. His feelings have been shared by coastal dwellers through the years. From the Paleo-Indians of many thousands of years ago to the booming resort population of today, all have been charmed by the lush, wild beauty and resources of the Georgia Sea Isles.

At least 12,000 years ago Paleo-Indians occupied the Georgia area. For ease of access to food and navigation, they usually established their communities near the banks of major rivers or at the coast near marshes and islands. Sea level at that time was approximately 250 feet lower than it is today, and the current continental shelf was an exposed coastal plain, probably occupied by the Paleo-Indians. As sea level rose at a rate close to 3 feet per century (until about 5,000 years before the present), the barrier islands and their inhabitants retreated landward. What was once coastal plain became flooded as part of the continental shelf. The Paleo-Indians changed their location and life-style with the rising sea level (behavior that today's coastal residents may be wise to emulate). Shell middens and other archaeological findings mark this pattern of settlements shifting with the changing environment.

The Native Americans known as the Guale, present when the first Europeans arrived, resided in the coastal area from north Florida to central South Carolina, and this region took their name. The last of the Guale disappeared about 300 years ago.

By the colonial era Creek tribes had established small villages on the islands and coastal areas. With the arrival of Europeans, however, first the French and then the Spanish, came the displacement and eventual destruction of these communities.

The French colonizing forces were eventually ousted by the Spanish; and the Spanish by the 1680s were pushed south into Florida by the formidable combination of attacks by Native Americans and English invasions from the north. Still, the English were unable to maintain a fort or settlement on the Georgia coast until the colony of Georgia was established by General James Oglethorpe in 1733.

The treaty of 1733 between Oglethorpe and the Creeks ceded all of the Georgia coast, except for Ossabaw, St. Catherines, and Sapelo Islands, to the English. These three islands, originally retained by the Creeks for hunting, fishing, and bathing, were relinquished to the English only one year later by a second treaty. When some of Oglethorpe's grand plans failed, lands in the new colony were deeded to English gentry by the king.

Subsequent occupation by English colonists resulted in the substantial modification of island environments. The islands continued to serve as important military vantage points; in addition, agriculture rapidly increased. On many islands large forests were cleared for crops. Sea Island cotton became a major product, along with rice and indigo.

On some of the islands, such as Cumberland, Sapelo, and St. Catherines, this growth was accompanied by the development of a Southern plantation culture, built on the toils of the growing slave population. With the advent of the Civil War, however, agricultural production on the island plantations and farms rapidly declined. By March 1862 all of the barrier islands had been seized by Union troops, and although some of the freed slaves began to rework island fields at war's end, the plantation era of Sea Island cotton had effectively ended. A few families that owned barrier island properties remained there, and some former slaves established small communities on Sapelo and Cumberland. Hog Hammock on Sapelo Island is the only such community remaining.

During the late 1800s the semitropical climate and wild beauty of the Georgia barriers lured new buyers to this economically depressed area. The purchase of Jekyll Island in 1886 by a group of wealthy industrialists heralded the era of exclusive island resorts. The Jekyll Club membership roster included such household names as Rockefeller, Goodyear, Pulitzer, Macy, and Morgan.

This new land ownership was characterized by single individuals or families acquiring entire islands or large portions of them. The majority of the islands were managed as private vacation retreats or homesites.

This pattern of ownership by a small number of families continued well into the 20th century. The present undeveloped and protected state of most of the coastal islands can be attributed to these earlier land-use and ownership patterns. Wassaw Island, for example, remained in the same family's hands from the 1860s to the 1970s, when it was sold and became a National Wildlife Refuge.

Difficult access and little demand for public use until the second half of this century also helped to preserve the islands in a relatively natural state. The development of any barrier island for recreational and, subsequently, for residential use depends on ease of access. As transportation to Tybee, Sea, St. Simons, and Jekyll Islands progressed from boats to railroads to automobiles, more people could reach the barriers with relative ease. The evolution of each of the four developed barriers of Georgia's coast has been determined largely by the nature and history of this public access.

Recreational use of Tybee Island began during the early 1800s when steamboat transportation made the island accessible to affluent Savannahians. With the formation of the Tybee Improvement Company in 1873, daily boat trips were scheduled, a hotel was built, and property was sold. The opening of a railroad route to Tybee in 1887 rapidly increased both recreational and residential use, and Tybee became the site of the first major modern island community on the Georgia coast. Construction of a highway to Tybee in 1933 opened it to even more people, spurring further growth.

During the early 20th century three other islands opened up to major public recreational and residential use. St. Simons and Sea Islands became attractive destinations with construction of causeways and bridges in 1924. St. Simons grew into a popular vacation and residential spot and in 1992 supports the largest residential population of any Georgia island. Meanwhile, Howard Coffin, the wealthy automaker who owned Sapelo Island and portions of St. Simons Island, began to develop an exclusive resort community on Sea Island. The character of present-day Sea Island is most similar to that of the seasonal resort of Georgia's recent past.

Jekyll Island was linked to the mainland by highway after the Jekyll Club sold the island to the state in 1947. Jekyll was designated a recreational area and presently allows limited residential use.

The growth of nearby coastal population centers has stimulated an increasing demand for ready access. Population growth on Tybee Island, near Savannah, and on St. Simons Island, near Brunswick, reflects the burgeoning of these mainland cities as well as easier access to the islands. Postwar prosperity and the automobile also have attracted more people, both Georgians and non-Georgians, from a much broader population base.

Development of the islands, particularly Tybee and St. Simons, was accompanied and sometimes encouraged by the growth of island infrastructure. Roadways, water supply, utilities, and services provide the framework that attracts more people to settle there.

Unfortunately, initial development of these four barriers occurred before people understood the nature of the physical processes that affect the islands; conflicts eventually arose between people's intentions and the islands' limitations.

Beach erosion and the typical response to it offer a prime example. Many island residents chose to protect their property from erosion by erecting seawalls and groins, often at the expense of the sandy beach. Today's challenge to Georgia's developed barriers is the development of appropriate, flexible, and economically and socially acceptable solutions to such conflicts.

Island Management: The Perspectives of the Present

Current island management strategies on use and development reflect the widely varied island histories. The Georgia coast has everything from New Jersey-style high-density development to virtual wilderness. Four islands serve as major residential and resort centers, while the remaining nine are, to varying degrees, presently protected from much coastal development.

Tybee and St. Simons Islands have been developed much like popular resort communities across the country. These two islands accommodate the majority of residents of the Georgia barriers, with the land largely owned in small parcels by individuals. However, the flavor of development on each island is slightly different. Although Tybee has experienced a flurry of new condominiums and beachfront houses, the island has retained the character of a seasonal beach community of the 1960s. St. Simons, a larger island with more space for sprawling development, has seen a rapid expansion of subdivisions as well as multiunit housing.

The response to beach migration on these islands has been typical. Initially piecemeal and eventually large-scale, publicly funded seawall construction has been the order of the day. Again, buildings are saved, beaches are lost (figs. 2.1 and 2.2). As a result, Tybee currently participates in a federal beach replenishment program, and a similar project is being considered for St. Simons.

Population pressures of the 1970s eventually prompted the development of land-use plans for each of the islands. Unfortunately, many problems affecting the developed islands began with the influx of residents in the early 1900s, long before any land-use plans were formulated.

Sea Island, another major residential center, has a much different atmosphere. Managed by the Sea Island Company, the exclusive resort community consists of expensive, single-family homes and small condominiums as well as the famous Cloister resort. In spite of the beauty and costliness of Sea Island's development, however, nature has not exempted the community from the effects of shoreline retreat. A wall of granite boulders or concrete protects the oceanfront houses (fig. 2.3); no beach exists at high tide; and the community is faced with inflating costs of trying to hold back the sea. The name "Seawall Island" might be the culmination of this approach. Sea Island has more recently begun its own beach replenishment program.

Jekyll Island, the other major residential recreational area, is owned by the state. To many people island management here has been par-

2.1 Seawall, no beach, on Tybee Island. Fort Screven area, October 1985. Photo by T. Clayton.

The Guale Coast: A Human Perspective 9

2.2 Seawall (riprap), no beach, on St. Simons Island. October 1985. Photo by T. Clayton.

ticularly disappointing. In spite of state ownership and control of development (managed by the Jekyll Island Authority), the island exemplifies unattractive (to some) and even unsafe development. Private homes, motels, and recreational facilities built in the 1950s and 1960s were not adequately set back from the ocean shore; when these structures were threatened by beach migration, the state allowed seawalls and revetments to be constructed for erosion protection (fig. 2.4). Recent development has been somewhat more sensitive to island environments and the erosion problem. The earlier shore-hardening structures, however, have already contributed to the loss of the beach. Although something has been learned from Jekyll Island's experience, it seems likely that public funds will continue to be used for erosion control, probably in the form of beach nourishment.

Although most coastal residents and visitors form their perception of the Georgia coast from visits to Tybee, Sea, St. Simons, and Jekyll Islands, an understanding of the remaining islands is important to a balanced perspective.

Little Tybee, Williamson, Wassaw, Ossabaw, St. Catherines, Blackbeard, Sapelo, Cabretta, Wolf, Little St. Simons, Little Cumberland, and Cumberland Islands are what we term the undeveloped islands of the Georgia barrier system. Almost all of them, as well as the relatively developed four, have experienced at least some degree of human use and alteration. In the course of subsistence use by Native Americans, early agricultural efforts by Spanish missionaries and Creek converts, and intense use during the plantation era, many island forests were cut and cleared. Timbering operations often accompanied the clearing of the land for crops. As mentioned, the Civil War devastated agricultural operations,

2.3 Seawall, no beach, on Sea Island. The house behind the wall is only a few (3 to 5) years old. February 1981. Photo by O. Pilkey.

10 Living with the Georgia Shore

2.4 Seawall, no beach, on Jekyll Island. Early 1980s. Photo by W. Cleary.

and the late 19th century saw wealthy families purchasing large parcels or entire islands. The last two decades have witnessed the transfer of many private properties to the public as parks, preserves, and wildlife refuges. Thus, with proper management, most of these islands should remain in their wild state.

The state manages Jekyll Island and four of these undeveloped islands. The northernmost, Little Tybee Island, was acquired in 1991 and will be managed as a wilderness preserve. Williamson Island is an ephemeral ribbon of sand subject to dramatic changes during winter storms. Little vegetation is found there, and no plans exist for any human use other than recreation. Ossabaw Island was sold to the state by the Torrey family in the 1970s. It has been retained in its undeveloped state as a wildlife preserve with minimal management. Sapelo and adjacent Cabretta Islands are used for wildlife management in the Game and Fish Commission refuge and for research and education at the University of Georgia's Marine Institute and the National Estuarine Sanctuary of the National Oceanic and Atmospheric Administration.

Most of the remaining undeveloped islands are managed by the federal government. Wassaw Island was sold in 1970 by the Parsons family to the U.S. Fish and Wildlife Service for a wildlife refuge. This island is the wildest on

Georgia's coast. Largely untouched by human modifications, it hosts a refuge with minimal habitat management for resident nongame species. Blackbeard Island, also a Fish and Wildlife refuge, has been similarly managed but with a slightly heavier hand and an emphasis on waterfowl management. Still, the island is relatively untouched. Wolf Island is also part of the Fish and Wildlife Service's coastal refuge system. Cumberland Island is a National Seashore of the National Park Service (NPS). Private landowners still hold large portions of Cumberland; however, these properties will become NPS lands on the deaths of the inholders, or the NPS will continue to purchase such lands until the entire island is federally protected.

Some undeveloped islands remain in private ownership. St. Catherines Island, managed by the St. Catherines Island Foundation of New York, is used for archaeological research, endangered animal research, and educational purposes. Little St. Simons Island is owned by a California family that manages the island as an exclusive nature resort. Little Cumberland Island, managed by the Little Cumberland Island Association, has 100 two-acre privately owned lots for limited and controlled development of single-unit dwellings.

The variety of owners and management strategies for these islands is unmatched elsewhere along "protected" stretches of the U.S. coast. In fact, the unique quality of the island ecosystems prompted the entire contingent to be nominated for a United Nations World Biosphere Reserve. Unfortunately, state officials successfully fought attempts to win this designation!

Although the undeveloped islands seem to be well-protected, the potential exists for their substantial degradation. Sale of private and possibly public holdings, or even subtle changes in management strategies, may adversely affect this unique system of wild barriers. At the present time all of them are held by owners sensitive to the uniqueness of the island environments. To one degree or another, these managers are committed to preserving the attributes that make the islands ecologically and aesthetically special and economically valuable. We hope their commitment continues.

3 The Guale Coast: A Natural Perspective

The natural features of the Guale coast (from southern South Carolina through northeastern Florida) differ in major ways from the remainder of the U.S. shoreline. Understanding the natural coastal system and the physical processes that shape it is important in determining where and how to best live "with" the Georgia coast, and especially with the islands.

The Coastal Setting

The great variety of natural beauty to be found in Georgia results in large part from three distinct physiographic provinces within the state's borders: the foothills of the Blue Ridge *mountains* of northwestern Georgia grade into the *piedmont* (literally, "foot of the mountains"), which falls off to the *coastal plain* in the east (fig. 3.1). The boundary between the piedmont and the coastal plain is known as the "fall line." This break marks the former ocean shoreline of millions of years ago during the Miocene Epoch.

Evidence of several other former shorelines can be found in the coastal plain within 120 miles of today's coast. Their locations explain why you can turn up sharks' teeth and seashells in places now high and dry (up to 100 feet above present sea level) and up to 60 miles inland.

These three physiographic provinces are interconnected by a system of rivers that 19th-century Georgians used as highways for transportation, trade, and communication (fig. 3.1). The Savannah River flows out of the Blue Ridge foothills and winds its way across the piedmont and coastal plain to the ocean. The Altamaha and Ogeechee Rivers originate within the piedmont, while such other rivers as the Satilla and St. Marys flow entirely within the coastal plain.

When these rivers reach the seaward edge of the coastal plain, they merge with a complex system of *estuaries* where extensive salt marshes up to five miles wide separate the mainland and barrier islands. In fact, Georgia and South Carolina have some of the largest areas of salt marsh in the United States. These marshes are among the world's most productive environments. They serve as important breeding grounds and habitat areas for a multitude of marine organisms, including many that are commercially important—crabs, shrimp, oysters, and fish.

Seaward of these marshes lies a chain of *barrier islands*. In serving as barriers that protect the mainland from the ocean's energies, these islands are highly dynamic environments. Sixteen islands make up the Georgia barrier system (fig. 1.2).

Beyond the barrier islands, to a depth of about 160 feet, lies the *continental shelf* (fig. 3.2). This broad, gently sloping shelf extends offshore for 70 to 80 miles. When sea level was much lower than today, the shelf was a forested coastal plain with gently flowing rivers. This lower stand of sea level (the most recent) occurred during the Ice Age when great amounts of water were locked up in glacial ice.

The position of the Georgia shoreline has ranged from the fall line, far inland, to the middle or even edge of the present continental shelf, far offshore. If you consider that the Atlantic Ocean, whose power and energy shape our coast, stretches for thousands of miles to Africa, you can see that the few *feet* of change we worry about at our beaches is relatively

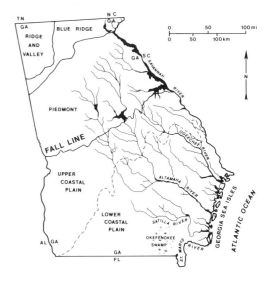

3.1 Georgia physiographic provinces. The boundary between the piedmont and the coastal plain is the "fall line," the approximate location of an ancient shoreline. (Adapted from *An Ecological Survey of the Coastal Region of Georgia*, by A. Johnson and others, 1974. (Complete reference given in appendix D under *Coastal Environments*.)

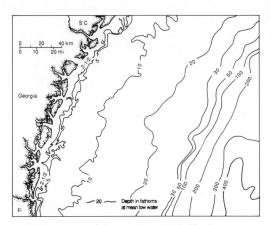

3.2 Bathymetry of the continental shelf adjacent to the Georgia coast. The numbers give the water depth in fathoms. (One fathom equals six feet). Even far offshore, the water is quite shallow. (Adapted from *An Ecological Survey of the Coastal Region of Georgia*, by A. Johnson and others, 1974. Complete reference given in appendix D under *Coastal Environments*.)

insignificant. And where we have made the decision to build right on that "thin edge," inevitable and expensive consequences will occur.

The Ups and Downs of Sea Level

Everyone who has visited the coast is familiar with the daily rise and fall of the tides, so a changing sea level is not an unfamiliar concept. In addition to these easily observed daily tidal changes, there are longer-term sea-level fluctuations, which are hardly noticeable to the casual summertime visitor but which can be measured with instruments or inferred from the geologic record.

If we measure sea level day in and day out for several years, we will find that the daily cycle of tidal change is actually superimposed on a longer-term cycle of changing sea level. Annual changes (or yearly cycles) in sea level (fig. 3.3a) can be produced by seasonal variations in climate and ocean water properties—such as temperature or atmospheric pressure.

Average sea level at a given place also can change significantly from year to year (fig. 3.3b) or decade to decade. For example, the sea level off southern California rose more than 2 inches during the 1960s—a significant increase over the preceding decade. This rise was coincident with a change in the prevailing wind pattern and a rise in water temperature.

Geologic evidence offers insight into sea-level changes of even greater magnitude (hundreds of feet) over longer periods. For example, about 20,000 years ago, during the last Ice Age, sea level was a few hundred feet *lower* than today (fig. 3.4). Conversely, about 100 million years ago, sea level was much *higher* than today, and much of the U.S. plains region was underwater. The geologic record of sea level shows a long history of dramatic and frequent ups-and-downs. These changes critically affect evolution of the continental shores, and of barrier islands in particular.

A list of factors that contribute to sea-level changes (table 3.1) shows that many agents are *local*, while others have *global* impact. For example, increased flow rates in the Savannah River will increase sea level in the Tybee Island area, but not in the state of Oregon. Melting polar icecaps, on the other hand, affect the world's oceans and contribute to a rise in sea level everywhere.

A record of sea level at any location will show the effects of such worldwide changes (called *eustatic* effects) in addition to the

3.3 (a) Average monthly sea level off the coast of California. (b) Average annual sea level on the California coast. (Adapted from "Variations in Sea Level on the Pacific Coast of the United States," by E. LeFond, in *Journal of Marine Research*, 1939, vol. 2, no. 1, pp. 17–29).

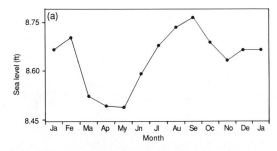

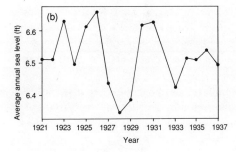

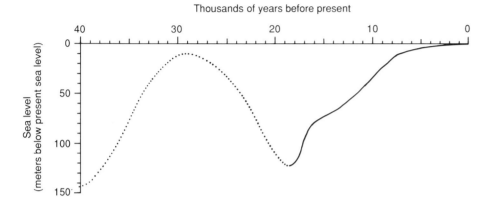

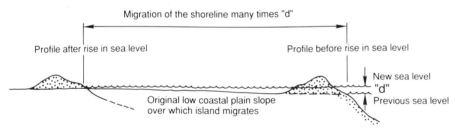

3.4 Fluctuations in sea level during the past 40,000 years. One meter equals approximately three feet. (Adapted from *Beach Processes and Sedimentation*, by P. Komar, 1974. Complete reference given in appendix D under *Beaches*.)

3.5 Relationship between vertical sea-level rise and horizontal shoreline movement (inundation plus erosion). On a gently sloping coastal plain, a small vertical rise in sea level results in a large landward shift of the shoreline; "d" represents the change in sea level.

superimposed imprint of local conditions. The result is *relative*, or local, sea level.

Eustatic sea level is rising at present and has been doing so for about 18,000 years (fig. 3.4) —since the glaciers of the last Ice Age began to melt away and pour their waters into the ocean basins. In Juneau, however, relative sea level is actually falling, since the land is lifting upward even faster than the seas are rising. In Savannah, on the other hand, the eustatic sea-level rise is augmented because the land is

Table 3.1 Some Factors Affecting Sea Level

	Global	Local
Volume of ocean basins (decreasing volume raises sea level)	X	
Volume of water in ocean basins (melting glaciers raises sea level)	X	
Water temperature (raising temperature raises sea level)	X	X
Vertical movement of the land (sinking the land is equivalent to raising sea level)		X
Local atmospheric pressure (decreasing atmospheric pressure raises sea level)		X
River runoff (increasing flow rate raises sea level)		X
Winds (onshore winds raise sea level)		X
Tides		X
Ocean currents		X

The Guale Coast: A Natural Perspective   15

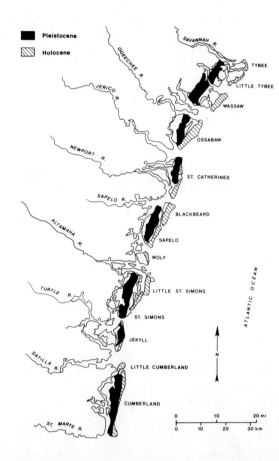

3.6 Recent (Holocene) barrier islands and older Pleistocene islands.

sinking; the rate of relative sea-level rise there is therefore greater than the rate of eustatic sea-level rise.

Along the U.S. East Coast the present rate of relative sea-level rise is about 1 foot per century. As in Savannah, this relative (local) rate results from the land sinking, plus the worldwide eustatic rise of the seas. On the Georgia–South Carolina coast the recent relative sea-level rise has averaged about 13 inches each century.

This vertical rise in sea level translates into a horizontal, or landward, change in shoreline position (fig. 3.5). Just how far and how fast this horizontal translation occurs is a complex function of sediment supply, rate of sea-level rise, slope of the land over which the sea is rising, and storm history. In general, the gentler the slope of the land, the farther the shoreline will migrate for a given increment of sea-level rise. Scientists estimate that on the Georgia–South Carolina shoreline a 1-foot vertical rise in sea level will result in the shoreline moving at least 200 feet landward, and considerably greater distances where the land is flat.

The largest and oldest of the current Georgia barrier islands were formed between 35,000 and 50,000 years ago, late in the Pleistocene epoch during a still-stand of the sea. The main parts of Cumberland, Jekyll, St. Simons, Sapelo, St. Catherines, and Ossabaw Islands are remnants from a time when sea level was 2 to 5 feet higher than today (fig. 3.6). Other barriers of this former system are Skidaway, Green, and Wilmington Islands, which were stranded as part of the coastal plain when sea level continued its decline some 25,000 years ago (fig. 3.4). The islands of Whitmarsh, Talahi, Oatland, Isle of Hope, Burnside, Colonels, Harris Neck, and Creighton are remnants of an even older former system. In fact, Georgia's coastal plain is marked by several older, former barrier island shores (fig. 3.7) that stand as high as 100 feet above today's sea level (table 3.2). These old barrier island sequences are now land-locked sand ridges, covered with scrub oak and pine.

Fifteen thousand to 18,000 years ago sea level was much lower than today (fig. 3.4), and the Georgia shoreline was many miles offshore. This occurred during the last Ice Age when vast glaciers covered the higher latitudes of the world, tying up great amounts of water.

Table 3.2 Georgia Barrier Island Sequences. Companion table to fig. 3.7.

	Sequence Name	Sea level at time of formation [feet (meters) above present sea level
I	Wicomico	95–102 (29–31)
II	Penholoway	69–75 (21–23)
III	Talbot	39–46 (12–14)
IV	Pamlico	23–26 (7–8)
V	Princess Anne	13 (4)
VI	Silver Bluff	3–6 (1–2)

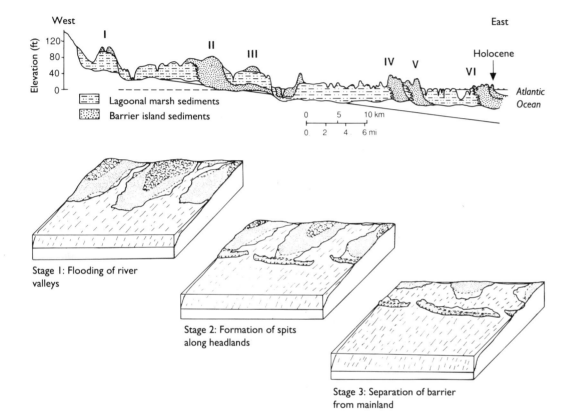

3.7 Diagrammatic cross-section of the lower coastal plain of Georgia, showing former barrier islands. Each barrier island system marks where the ocean shoreline was in the past, when sea level was higher than it is today. For more information, see Table 3.2. (Adapted from "An Appraisal of the Evolution of the Lower Coastal Plain of Georgia," by J. Hails and J. Hoyt, 1969, *U.S.A. Trans. Inst. British Geog.*, vol. 44, pp. 133–139.)

3.8 One scenario for the formation of barrier islands. Bays develop first as sea level rises. Then spits form from sand delivered by erosion of the headlands between the bays. Finally, rising seas flood the land behind the spits and isolate them as barrier islands.

As the ice started to melt, the sea began a rapid rise (about 3 feet per century). This water flooded the river valleys, forming *embayments*, or *estuaries* (fig. 3.8). If you look at a map of today's shoreline, you can see many such inundated valleys, especially along the U.S. Atlantic Coast. Chesapeake Bay and Delaware Bay are prominent examples.

If this were all that had happened, the shoreline today would be jagged. Nature, however, tends to straighten irregular shorelines. Wave action cut back the *headlands*—the ridges of land that extended seaward between flooded valleys—and built *spits* extending from the headlands across the bay mouths. As sea level continued to rise, the low-lying land behind the spits was flooded. This flooding behind the old dune and beach line, plus spit breakthroughs that created inlets, resulted in spits separating from the mainland. The barrier islands were born. The precursor of present-day Tybee Island, for example, was created in this way, far offshore from where the barrier is today. This is one of several ways that barrier islands are formed.

As sea level continued to rise, the barriers responded by moving or migrating toward the mainland (fig. 3.9). The more rapidly sea level rose, the more rapidly the barrier islands migrated across the continental shelf. (Today this is happening dramatically in Louisiana.) The mainland shoreline was also retreating, flooded by the rising seas. Eventually, the rising waters began to flood around the coastal

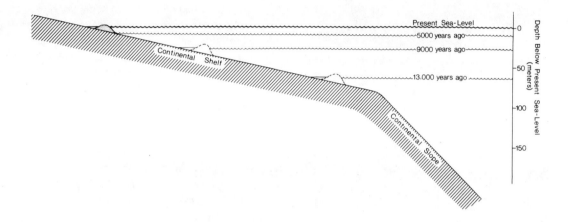

3.9 Landward movement of Holocene barrier islands across the continental shelf as sea level rose over approximately the last 15,000 years. (One meter equals approximately three feet).

plain remnants of the former Pleistocene barriers. As these remnants were surrounded by ocean to again become islands, and the Holocene barriers migrated landward, the present coastline began to take shape.

Some of the Holocene barriers migrated into, welded on, and wrapped around the Pleistocene islands (fig. 3.6). The combination of these two island systems explains why the Georgia islands are among the widest in the world. In fact, *sea island* in the geologic literature means "wide barrier island."

Some of the second generation (Holocene) islands, influenced by nearby rivers, did not merge with their older counterparts. The sediment carried by rivers like the Savannah and Altamaha formed large deltas at their mouths. This input of sediment into the coastal system tended to counteract the effects of rising sea level and slowed landward migration of the Holocene islands immediately south of the river mouths. The back-barrier area of the deltas became the vast salt marshes of today. One example is the Tybee-Little Tybee-Williamson Island complex of the Savannah River "delta" (fig. 3.6).

The evolution of the shoreline continues, of course. About 5,000 years ago the rate of sea-level rise slowed considerably (fig. 3.4), resulting in a relatively slow rate of shoreline migration. This period of relative stability, however, may have ended recently. About 50 years ago the rate of relative sea-level rise apparently began to quicken. A rise in global atmospheric temperature, this time prompted by human activities, also may be occurring. If it is, we can expect the oceans to warm and global (eustatic) sea level to rise. Accordingly, the rates of shoreline evolution and change may be increasing, with interesting and potentially devastating results.

The Greenhouse Effect and Sea-Level Rise

The retention of solar energy within the earth's atmosphere by water vapor, carbon dioxide (CO_2), and other naturally occurring atmospheric gases accounts for the habitable climate on earth. Without the action of these "greenhouse gases" in trapping heat, the earth would be a frigid planet, with an average temperature approximately 30° Celsius below present averages.

Since the beginning of the industrial age (mid-1800s), atmospheric levels of carbon dioxide, methane, nitrous oxide, and chlorofluorocarbons (CFCs) have dramatically increased. The combustion of fossil fuels, deforestation, and other human activities are responsible for these rapid changes. Given current rates of CO_2 production and introduction into the atmosphere, scientists expect an effective doubling of carbon dioxide levels by the year 2050. As a result, global atmospheric temperatures could rise between 1.5 and 4.5 degrees Celsius as more of the sun's radiant heat is trapped within the earth's atmosphere; the best guess is 2.5 degrees Celsius. A host of major environmental changes would result from such a significant temperature increase, and many social problems would follow.

Global sea level, for example, is of significant concern. Among the results of global warming would be (1) expansion of waters already in the ocean basins—just as any water expands when heated (called *thermal expansion*), and (2) increased melting of glaciers and polar ice sheets, releasing even more water into the ocean basins. Both conditions would contribute to a rapid sea-level rise, much as at the end of the last Ice Age. The current estimate is a rise of 3–11 inches by the year 2040.

Although some media coverage has questioned whether "greenhouse warming" will occur, an overwhelming majority of scientists (specifically, climate change experts) do expect significant global warming to occur. The exact nature of that change is uncertain. With continued research, predictions of potential changes are frequently revised. To obtain the most recent information, consult the research literature and active researchers. For an excellent summary of the scientific consensus as reported in 1990, see the reference by Richard A. Kerr (appendix D under *Sea-Level Changes*).

The safest assumption you can make about future sea-level rise is that it will continue and accelerate. The National Academy of Sciences has warned that much evidence points to a warming of the earth's surface; the Intergovernmental Panel on Climate Change concurs. If such dire predictions turn out to be accurate, we may no longer be able to afford protection of beachfront condominiums and tourist amenities. Our society's attention and financial resources will instead be focused on protecting the shores of our major coastal cities—such as Miami, Charleston, and New York—from flooding.

Barrier Island Processes: An Integrated System

Changes in island shape and position are caused by several related physical processes. Sea-level changes, waves, tides, currents, winds, and storms are the major physical forces highlighted in the following section. Long-term and short-term, cyclical and catastrophic, such coastal processes continuously sculpt the changing face of the Georgia islands.

Beaches: Their Dynamic Equilibrium

The beach is one of the earth's most dynamic environments. The beach—or zone of active sand movement—is always changing and always migrating. Keep in mind that the beach extends offshore from the toe of the dune to a depth of 30 to 40 feet. The intertidal and exposed part on which we walk is only the uppermost portion of this system.

The condition of the beach system depends on three major elements: wave energy, sea level, and sand supply. The complex interaction and balance among them determine both

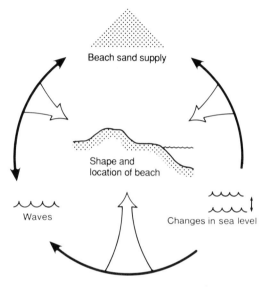

3.10 The dynamic equilibrium of the beach.

(1) the *shape* of the beach and (2) its *position*. The natural balance engendered by this relationship of factors is referred to as *dynamic equilibrium* (fig. 3.10). When one factor changes, the others adjust accordingly to maintain a balance. When we enter the system incorrectly—as we often do—the dynamic equilibrium continues to function but in a way that may not be to our liking.

Origin of Sands of the Shoreline

The sands of the coastal plain, barrier islands, and nearshore continental shelf regions were originally derived from weathering of the piedmont and the Appalachian Mountains. Today, however, rivers bring relatively little sediment to the ocean shores along most of the eastern U.S. coast.

Instead, river-borne sand is typically deposited far inland at the heads of estuaries, behind man-made dams such as those at the Hartwell and Clark Hill Reservoirs in Georgia, or in unnaturally deep dredged channels. Soil conservation measures also decrease the sediment load of coastal rivers.

Erosion of the coast itself also provides some beach sand. For example, wave erosion of cliffs of glacial sediments provides most of the sand to the beaches of Long Island.

Along most of the U.S. barrier coast—which runs approximately 10,000 miles from the south shore of Long Island, New York, down and around to where the Texas coast meets Mexico (with a small break in the "armpit" of Florida)—much of the beach sand comes from the adjacent continental shelf. It is pushed up to the beach by fair-weather waves.

Additional sand, sometimes extremely large quantities of it, is carried alongshore in the surf and intertidal zones. This sand reflects no additional sediment input to the overall coastal system; longshore transport merely redistributes what is already there. Locally, however, longshore transport is often the immediate source of most of a beach's sand.

In some areas, beach sediments may also be composed of *biogenic* materials—that is, materials of biological origin, such as coral fragments or mussel shells. In mid-latitudes, however, beaches are generally less than 25 percent biogenic debris.

It is important for beach dwellers to know about, or at least have some sense of, the source of sand for their beach. The dredging of navigation channels and the construction of groins may reduce the transport of sand alongshore to adjacent islands or beaches, especially those to the south (on the Atlantic coast). For example, the dredging of the Savannah River to the north of Tybee Island is thought to contribute to erosion problems on Tybee.

Community actions taken on an adjacent island or inlet could affect your beach, just as your own actions may affect your coastal neighbors.

Moving Sand: Wind, Waves, and Currents

The interplay of wind, waves, and currents determines the direction and rate of sand transport. Knowledge of the seasonal and cyclical fluctuations of these agents of change is essential to understanding the patterns of "short-term" changes on the barrier islands.

Winds and Waves

Most ocean waves are generated by winds blowing across the water's surface. In fact, most waves that strike the shore are produced by large storms far out to sea.

In the summer (late spring through early fall) the Georgia coast generally enjoys soft southeasterly and easterly winds accompanied by gentle ocean swell. These swell waves generally transport sand toward the shore, widening beaches, and, in conjunction with onshore winds, augmenting the dune system. Hurricanes and local summertime storms occasionally will interrupt the beach-building, but only temporarily.

In the winter the dominant agents of change are the high winds and accompanying steep waves of northeasterly storms. Annual net sand transport from north to south is caused by this energetic wave action.

On the Georgia coast the average wave height is only about 1 foot, partly because of the damping effects of the broad continental shelf (fig. 3.2). The gentle (2 feet per mile) slope of the shelf results in much energy of the approaching swell being dissipated before the waves actually strike the shoreline. Average wave height is lowest near the center of the Georgia Bight (near Savannah) and increases to the north and south where the shelf narrows and the grade of its slope steepens. This relatively low wave energy partly accounts for the relatively low rates of shoreline recession observed in Georgia.

Waves generate longshore currents that flow parallel to the beach (fig. 3.11). These currents are familiar to anyone who has swum in the ocean; they are the reason you sometimes end up somewhere down the beach, far away from

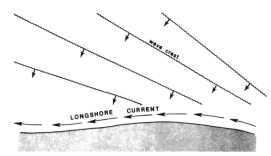

3.11 Overhead view of beach and nearshore region. When waves break at an angle to the beach, longshore currents flowing parallel to the shore are generated.

your towel. Such currents result when waves strike the beach at an angle, and a portion of the breaking wave's energy is directed along the beach. These currents are capable of carrying large amounts of coarse material for miles along the beach, especially along straight coastlines with gently sloping beaches. The sand thus transported may be deposited at the end of the island—hence the spit growth seen on the south ends of several Georgia islands (for example, Blackbeard and Sea Islands). Longshore currents may also be associated with the development of rip currents—strong, narrow currents that flow seaward from the surf zone.

This longshore transport has been likened to a "river of sand" by which sand is transported continuously along beaches and from island to island on most Atlantic and Gulf barrier coasts. However, some geologists believe that the river-of-sand concept doesn't adequately describe the movement of sand *between* island units on the Georgia coast; it works only *within* units, such as Blackbeard-Cabretta-Sapelo. According to the "semi-enclosed sediment cell concept," the eight major island units (or groups) (fig. 3.12) are isolated from adjacent units by relatively deep inlets and large-volume tidal flow. The inlets act to define each end of a cell within which sediment is constantly reworked but only infrequently removed for transport farther south. Significant southward sand movement out of the sediment cell is thought to occur only during major storms.

Tides

Two high tides and two low tides occur daily on the Georgia coast. Twice-daily tides with approximately equal highs and lows, are known as *semidiurnal* and are experienced all along the Atlantic coasts of North America and Europe, with about 12 hours and 25 minutes between each low tide.

The motions of the earth, moon, and sun control the tide's amplitude (size) and periodicity (timing). When the sun and moon are aligned with the earth, during full moon and new moon phases (fig. 3.13a, top), the largest tidal ranges occur. These are *spring* tides (fig. 3.13b). The highest spring tides occur about three days after the new moon.

When the alignment of moon, sun, and earth

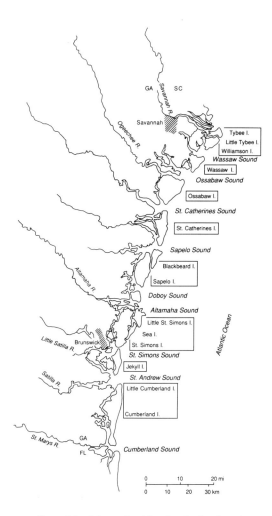

3.12 Georgia's eight major island units (each unit enclosed by a box).

The Guale Coast: A Natural Perspective **21**

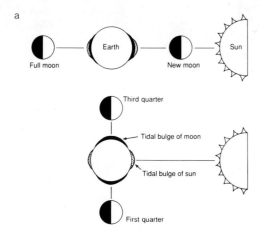

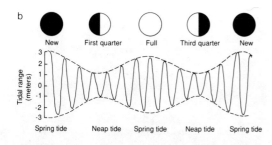

3.13 (a) and (b) Spring tides are produced during the full and new moons. Neap tides are produced during the first and third quarters of the moon. (Adapted from *At the Sea's Edge*, by W. Fox, 1983. Complete reference given in appendix D under *Coastal Environments*.)

creates a right angle, during the first and third quarters of the moon (fig. 3.13a, bottom), *neap* tides occur (fig. 3.13). During this interval the difference between high and low tides is at a minimum.

In midocean, tidal ranges are quite small—only 2 feet or so. Nearer the coast the magnitude of the tidal range increases; the broad slope and relatively shallow water of the continental shelf help to build up the higher ranges. In addition, the arc of the regional embayment (Georgia Bight) "funnels" tidal waters into its central area, creating the greatest tidal ranges at this point, near Savannah, with decreasing ranges to the north and south. These large tides, typically 6 to 10 feet in Georgia, are in part responsible for the coast's tremendous intertidal areas—the vastly productive salt marshes.

The ebb and flow of the tides create tidal currents, important agents of sediment transport—especially near inlet areas. An *inlet* is the channel of water between adjacent barrier islands. It may be a permanent feature at a river or stream mouth, or it may be short-lived, formed when water breaches the island during a hurricane or major storm. The permanent inlet type is more common in Georgia.

Two major sand bodies are typically associated with a tidal inlet (fig. 3.14): a *flood tidal delta* and an *ebb tidal delta*. The flood tidal delta, found on the landward side of the inlet, is formed and shaped by the currents of the incoming (flooding) tide. The ebb tidal delta, which forms on the ocean side of the inlet (figs. 3.14 and 3.15), is made up of sands being carried out of the estuary; it is shaped by the interaction of the currents of the falling (ebbing) tide and longshore currents generated by waves. Georgia is world-famous for having extremely large ebb tidal deltas, extending up to 4 miles offshore.

Tidal currents can interact with waves to bring about shoreline changes as well. For example, the recurved spits at the south ends of

3.14 Tidal deltas form at inlets between islands. In Georgia the ebb tidal delta tends to be much larger than the flood tidal delta. Stippled areas indicate shoals; dashed lines outline areas of greater water depth.

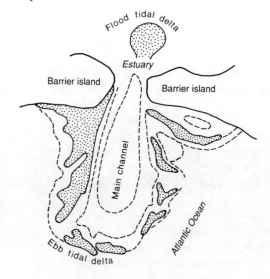

3.15 Ebb tidal delta at the Tybee Creek inlet between Tybee Island and Little Tybee Island, 1987. Photo by L. Taylor.

some Georgia islands are produced by this interaction, Jekyll Island being one example. Flooding (incoming) tidal currents also may carry sand from oceanfront beaches to riverfront beaches at the north ends of islands, as at Tybee Island. The ebb tidal delta shoals also constitute an important reservoir of sand, some of which is periodically brought ashore as waves induce shoal migration toward the beach.

Erosion—Why and Where Does the Sand Go?

Beach erosion occurs when more sand is carried away from an area than is replaced during a given period; therefore, erosion may be exacerbated by either increasing sand transport away from the beach or by diminishing the supply of sand coming to the beach.

Most beach erosion on the Georgia coast occurs from late autumn through spring. Increased wave energy, largely the result of passing northeasterly storms, is responsible for transporting large volumes of sand alongshore and offshore. Reductions in sediment supply also may contribute to erosion, as seen on Tybee Island (channel dredging, groin construction) and on St. Catherines Island (far from the source of river sediment).

Rising sea level, however, is the longer-term driving force for shoreline recession. The rate of sea-level rise is presently about 1 foot per century.

Martha Griffin and Vernon Henry, two Georgia geologists, have stated that "this relatively gradual rise is the single most important long-term agent of shoreline erosion." Georgians can be thankful not to have the 4-feet-per-century sea-level rise of Terreborne Parrish in Louisiana. As mentioned, rates of sea-level rise can be different in different coastal areas because the land may be slowly sinking or rising relative to worldwide ocean levels.

Working in conjunction with the rise in sea level are the many forces of nature, including waves, tides, and wind. These forces, which move sand about, are aided by the effects that people create: jetties and groins, or dams on rivers, for example. Any process that affects the amount of sand arriving on or leaving from a beach will affect the erosion rate.

Erosion rates vary along the Georgia coast in response to these factors. Generally, the northern and northeastern ends of the islands are more often subject to erosion. However, some islands experience cycles of accretion in these areas, as when ebb tidal shoals are moved shoreward and added to the beach. The southeastern tips of some islands are typically subject to growth through accretion; however, some islands exhibit long-term trends of erosion in these areas. In regard to patterns of erosion, each island must be considered as an individual case.

The study by Griffin and Henry of mean high water shoreline changes from 1857 through 1982 (cited in appendix D) asserts that "no discernible net shoreline erosion has occurred on the Georgia coast" during those years. This does not mean that no erosion occurred, but that from 1924 through 1974 a dynamic stability was maintained within the system of all Georgia barriers. For instance, erosional trends on Tybee and St. Catherines Islands were balanced by trends of accretion on Little St. Simons and south Cumberland Islands. A dynamic equilibrium also was maintained on some individual barriers. Many of the islands have been reshaped, without their migrating to another location. Of course, erosion along even a small portion of an island shoreline can be a potentially severe problem for residents.

Erosion is not limited to ocean and inlet shorelines. Shoreline erosion is also a natural event on the back side of islands along sounds and tidal creeks. Especially for upland areas on the outer edge of the bend of a river or creek, erosion may be a problem. However, erosion rates haven't been determined for most of these back side areas.

How to Recognize an Eroding Shoreline

Several clues can indicate a trend of erosion along a beach. Sand dunes that have been cut in half (or truncated), upland scarps, tree stumps on the beach, exposed deposits of marsh clays and peats among beach sands (fig. 3.16)—these are indicators of recent beach erosion. Eroded sand dunes may be evidence of recent, perhaps only seasonal, erosion. The exposure on the beach of upland or back-barrier features, however, may indicate significant and perhaps long-term erosion back into forests or over marshes. An eroding shore will migrate back over such features, often exposing long-buried surprises.

On Wassaw Island the skeleton of a dolphin and the burned-out hull of a wooden boat were found embedded in old marsh mud on the eroding north end beach (fig. 3.17). The boat and dolphin had washed into the marsh many years ago. Today, the north end beach, which is receding at a rate of about 15 feet a year, has migrated into and over the old marsh, uncovering these remnants.

Seashells also offer clues about shoreline migration. Many shells along the shoreline may be fossils, especially in such areas of erosion. Radiocarbon dating of shells from Atlantic beaches reveals that some shells are thousands of years old, which is what we would expect, if they are indeed from former backside areas now exposed on the front side. On migrating shorelines, you also can find shells of organisms such as oysters and snails whose natural habitat is the quiet marsh environment, not the high-energy, turbulent beach zone. Such shells may be washed up from old deposits of marsh clays that are exposed as the island shoreline migrates back over the former marsh.

Of course, many shells along the shore are not fossils. These treasures of the sea are former homes of marine mollusks that live, often for decades, in or on the sands of subtidal nearshore areas.

For many shell collectors the Georgia beaches seem to be a vast wasteland. The small number of shells on island beaches leads some to believe that few animals live offshore. Not so! Unless large, strong waves strike a shoreline, heavy shell material cannot be carried to intertidal beaches. The best time to collect shells is after or during a coastal storm from late autumn until early spring. The large waves associated with such storms often bring thousands of shells (usually with animals inside) to the beaches.

Tree stumps on the beach also provide evidence that shoreline migration is occurring (fig. 3.16). The trees certainly never grew on the beach. They once were part of the interior maritime forest or grew near the backside of the island; now the island shoreline has moved, and the beach has caught up with them.

Accretion—The Beach Widens

Accretion, the counterpart to erosion, occurs when more sand is deposited along a section of shoreline than is transported away. This process spurs the growth of the beach and eventually the island. The transport of sands from eroding to accreting areas sets in motion the migration of an island shoreline and eventually changes the island's shape and location.

Beaches may grow by different methods and in various directions through accretion. Often sand comes ashore from the continental shelf by the incorporation of ridges that form a few tens of yards off the beach. These sand ridges tend to slowly move onshore, a few feet per day, until they become part of the beach.

The next time you're at the beach, observe the offshore ridge (where waves may be breaking) for a few days and verify this for yourself. You may find that each day you have to swim out a slightly shorter distance to stand on the sand bar.

At low tide during the summer the lower beach frequently has a trough filled or partly filled with water. This trough is formed by a ridge that is in the final stages of "welding" onto the beach. Several ridges combine to make the berm, or beach terrace, on which sunbathers loll.

The welding of a lobe of sand from an inlet shoal, either through storm action or through more gradual migration of a tidal inlet channel away from the beach, also results in rapid

3.16 Signs of shoreline recession: (a) Erosional scarp, Sea Island. Early 1980s. Photo by W. Cleary. (b) Trees and stumps on the beach, Wassaw Island. Photo by O. Pilkey. (c) Outcrops of marsh deposits (mud, peat) on the beach, Sea Island. Seawall and revetment in the background are also evidence of erosion. Photo by W. Neal.

growth of the beach. In Georgia this tends to occur at the northern ends of the islands.

Southward sand transport by longshore currents may result in the elongation of a spit of an island's south end. Blackbeard Island and Sea Island have excellent examples of such spits. The Sea Island spit grew about 1 mile during just the past 120 years (fig. 3.18)! The southward growth of five of the Georgia island units is due to such patterns of accretion.

3.17 Hull of a boat and marsh deposits uncovered by shoreline recession at the north end of Wassaw Island. 1985. Photo by L. Taylor.

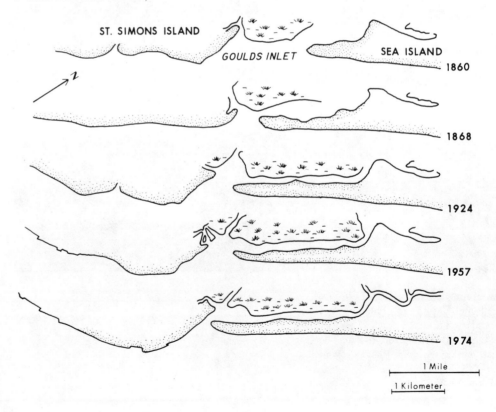

3.18 Migration of Goulds Inlet and elongation of the Sea Island spit, 1860–1974. (Adapted from *Historical Changes in the Mean High Water Shoreline of Georgia, 1857–1982*, by M. Griffin and V. Henry, 1984. Complete reference given in appendix D under *Individual Islands*.)

26   Living with the Georgia Shore

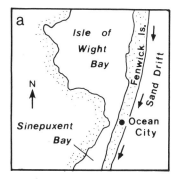 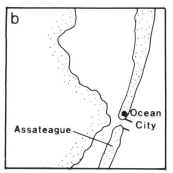 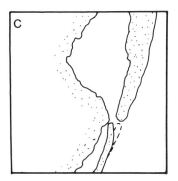

3.19 A modern example of barrier island migration: Assateague Island, Virginia. (a) Before 1933 Assateague Island, Virginia, and Fenwick Island, Maryland, were one continuous island. (b) A 1933 hurricane created Ocean City Inlet, separating the two islands. Subsequently, the U.S. Army Corps of Engineers installed two jetties there to keep open the new inlet. (c) The jetties disrupted the natural sand transport system, resulting in extensive accretion (sand buildup) on the north side of the inlet and in severe erosion and island migration on the south side. (Adapted from *Report of the Barrier Island Work Group*, U.S. Department of the Interior, Heritage Conservation and Recreation Service.)

Island Migration

Erosion in one area of island, and accretion in another, eventually lead to migration of the island as a whole. We can directly observe the migration of some entire barrier islands. The long jetties at Ocean City Inlet in Maryland, for example, have cut off the supply of sand coming to Assateague Island, just to their south. Cutting off the sand supply has the same effect as a rising sea level. Since 1935, when the inlet jetties were built by the Corps of Engineers, we have been able to watch by means of aerial photographs the migration of Assateague Island to a position completely behind where it was just 57 years ago (fig. 3.19). While these rapid changes have helped scientists studying island migration, they have created severe problems for the National Park Service, which owns Assateague, and for the Corps of Engineers, which is responsible for keeping the island from migrating back into the Atlantic Intracoastal Waterway.

In Georgia such front-to-back rolling over of barrier islands has not been prevalent in the recent past. Instead, some Georgia islands more commonly shift so as to rotate counterclockwise. Three Georgia islands have migrated southward in the past 120 years.

Storms and Beaches—The Sand-Sharing System in Action

Visitors to barrier islands are often surprised at how flat and broad the beach is after a storm. The gently sloping, relatively flat post-storm beach can be explained in terms of dynamic equilibrium (fig. 3.10): as wave energy (steepness) increases, the sands of the upper beach and dunes are eroded and transported to the lower beach or offshore, changing the beach's shape (fig. 3.20). As the beach flattens, storm waves expend their energy over a larger area. Typically, an offshore bar (ridge of sand) also forms, and larger storm waves will "trip" on this bar before reaching the beach.

The combined result is that the waves striking the beach are smaller and less damaging than if the beach had not flattened or not formed an offshore bar. However, since the waves take sand from the upper beach and transport it to the lower beach, a cottage located on the first dune may disappear along with the dune sands.

The sand bar produced by storms is easily visible during calm weather as a line of surf a few tens of yards off the beach. Geologists refer to the bar as a *ridge* and the intervening trough as a *runnel*.

During a storm an island can "lose" a great deal of sand to submerged shoals and sand bodies. Much of it will eventually come back, gradually pushed shoreward by fair-weather swell. After waves return the sand to the

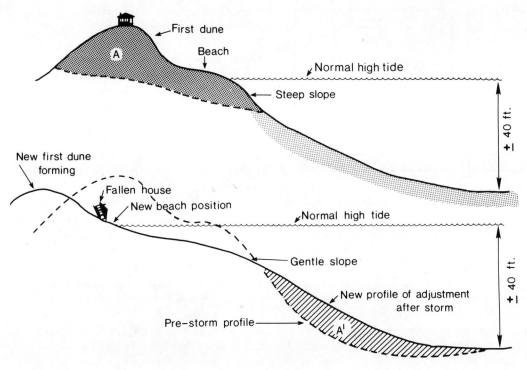

(Shaded area A' is approximately equal to shaded area A.)

3.20 Beach flattening in response to a storm.

beach, the wind takes over and slowly rebuilds the dunes, storing sand to respond to the next storm. Return of the beach may take weeks, months, or even years. If the storm waves carry the sand too far offshore, beyond the reach of recuperative fair-weather waves, some of it may never return. (During a storm there may also be net gains or losses due to alongshore transport of sand.)

The beauty of the sand-sharing system is that dynamic equilibrium allows the island to adjust to and recover from the erosive effects of storms. Changes in shoreline position or island shape may result; such changes are inevitable. Yet the shore manages to maintain the dynamic equilibrium that allows the island to function as a natural unit. The beach, for example, may change its shape or position, but its existence is not threatened.

Developed shorelines, on the other hand, may not be able to react to storms in this manner because the dynamic equilibrium of the sand-sharing system has been disturbed. Seawalls, for example, isolate the beach from the reservoir of sand stored in the dunes, thus disrupting the beach system's natural storm response.

Coastal Storms—Hazards, Headlines, and Heartbreaks

The Georgia coast is subject to two kinds of significant coastal storms: (1) *hurricanes*, and (2) *northeasters*. Hurricanes, with their accompanying high winds and waters, pose the greatest threat to the lives of coastal residents. Northeasters, on the other hand, which occur more frequently and are of longer duration, probably do more of the beach erosion over the long term.

Nor'easters

Devastating hurricanes such as Georgia's Great Hurricanes of 1893 and 1898, the Gulf's Camille of 1969, and South Carolina's Hugo of 1989 grab most of the media coverage and public attention. News coverage and public awareness is much lower for the storms that actually cause most of Georgia's annual net beach erosion—the *northeasters*.

During the fall and winter months along the Atlantic coast, large, intense low-pressure systems develop and produce high winds and waves that approach the shoreline from the northeast. These nor'easters tend to hang around and pound the shore for days, transporting tremendous volumes of sediment offshore and southward along the shore.

Occasionally, these storms coincide with the spring tides (the twice-monthly occurrences of the highest tides); at these times storm waves are superimposed on the extra-high tides and produce intense periods of erosion. In October 1981, for example, Georgia experienced its highest astronomical tides in more than a decade. A mid-October northeaster, assisted by the high tides, crumbled more than 450 feet of an "ideally" designed seawall on Sea Island (fig. 3.21). The storm's effects also were felt elsewhere along the Georgia coast.

This scene was relived along the U.S. East Coast during the New Year's Day storm of 1987 when a northeaster again coincided with a high tide. On Tybee Island flooding occurred along much of the shore. The south end parking areas and adjacent streets were completely flooded over a four-block area.

The effects of these storms, however, are dwarfed by those of the March 1962 Ash Wednesday Storm, one of the worst recorded in the history of the U.S. Atlantic shore. This late-winter storm struck during a spring tide and lingered offshore for three days. Shoreline

3.21 Sea Island seawall after passage of 1981 northeaster. Photo by L. Taylor.

erosion was reported from Florida to Massachusetts, with the hardest-hit areas in New York and New Jersey. Entire islands were topped by walls of water 20 feet high. Storm waters piled high in the lagoons and estuaries, causing massive flooding. New inlets were created. Even a Navy destroyer was washed ashore at Harvey Cedars, New Jersey.

The occurrence of such storms back-to-back is not uncommon, and it is one reason northeasters can be particularly damaging. In such cases the beach has no time to build up naturally during the short interval between storms. The North Carolina islands, for example, suffered much damage in December 1986 when the beach responded naturally to the passing northeaster by moving sand offshore; then, before homeowners had recovered, the coast was struck by the spring tide and the superimposed northeaster of January 1987.

Fortunately, Georgia did not experience major damage from either of these storms. Yet, a year rarely passes that the state's coastline does not experience some erosion from northeasters. Newspapers document the effects of these storms on human property over many decades. Failed seawalls, collapsed bulkheads, eroded dunes, and demolished buildings testify to the power of these storms.

Overwash of the primary dune line can occur during strong storms. Fans of overwashed sand may be deposited behind breaches in the dunes (fig. 3.22). In some areas, like Texas, storm waters may wash across the entire

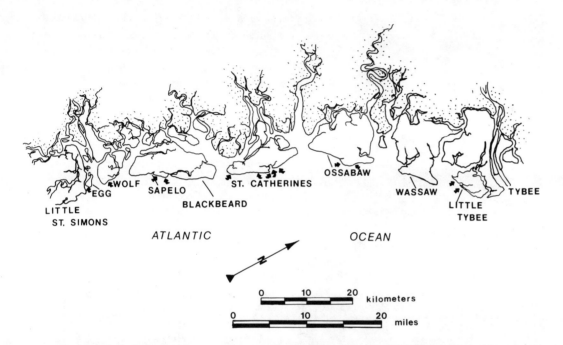

3.22 Washover activity on the Georgia coast. Arrows point to locations of washover fans as of 1976. (Adapted from "Holocene Depositional Environments on the Georgia Coast and Continental Shelf," by J. Howard and R. Frey, 1980, in *Excursions in Southeastern Geology: The Archeology-Geology of the Georgia Coast*. Complete reference given in appendix D under *History*.)

island, depositing sand in the backside lagoon. However, because of the relatively large size of the Georgia islands, most barriers are never entirely overwashed except on low spits or by major hurricanes.

Williamson Island is an exception. It is subject to major changes wrought annually by northeasters. Completely overwashed by storm waters, the scattered clumps of island vegetation are usually swept away with much of the sands as shoreline position and island configuration change dramatically. Williamson's highly dynamic nature is more typical of islands of other coastlines with smaller tides and larger waves.

Predicting when and how many of these storms will strike is impossible. What we do know is that they *will* strike and that their effects are predictable. Northeasters on a natural shoreline result in shoreline recession, with at least some subsequent recovery. Northeasters on a walled shoreline result in vertical erosion and rapid disappearance of the beach, with perhaps reduced subsequent recovery. Potential damages from these storms can be reduced by ensuring that no structures are located too near the beach, especially on islands without a healthy dune system.

Hurricanes: Killers at the Coast

Sea-level rise may be the ultimate villain in the shoreline saga, but hurricanes are the most memorable actors. Wind, waves, storm surge (the increase in water level during a storm), and overwash reach their peaks during hurricanes, but the majority of today's coastal residents and property owners have never experienced one. The relatively hurricane-free period from the 1960s through the 1980s has contributed to an apathetic disregard of the hurricane menace and increased development in high-hazard zones. However, time is not on the side of such development, as Hurricane Hugo demonstrated in South Carolina in 1989.

On June 1 of each year the official hurricane season begins. For the next six to seven months conditions favorable to hurricane formation can develop over the tropical to subtropical ocean waters of the Western Hemisphere. Hurricanes that ultimately strike the eastern United States tend to originate in the Gulf of Mexico or the Caribbean Sea early in the season, or, more likely, form later in the season (August, September, and October) in the eastern North Atlantic Ocean. Georgia hurricanes have struck as late as October and as early as May, but they occur most often in August (table 3.3).

Although meterologists are still seeking answers to the causes and mechanics of hurricanes, the basic model of what happens is known. During the summer the surface waters off West Africa heat up to at least 79°F. Evaporation produces a layer of warm, moist air over the ocean. This moist air is trapped by warm air coming from the African continent, but some is drawn upward. As the moist air rises, it cools and condenses, releasing heat, which in turn warms the surrounding air and causes it to rise. As a result, a low-pressure area forms (tropical depression), and warm easterly winds rush in to replace the rising air. The effect of the earth's rotation deflects the air flow, and the counterclockwise (clockwise in the Southern Hemisphere) rotation of the air mass produces the familiar shape of a hurricane. Air forced to the middle of the spiral can move only upward, producing a chimneylike column of rising air—the eye of the storm.

So a heat engine effect evolves with rising moist air cooling and condensing, releasing heat to cause more air to rise, allowing more air to rush in over the sea, a seemingly endless source of moisture. Heavy rainfall characterizes the edges of the cloud mass. When sustained winds reach 74 mph the storm is classified as a hurricane. The strongest winds of a hurricane may exceed 200 mph, but the actual maximum winds that have hit coastal areas are generally unknown because wind-measuring instruments are blown away.

Once formed, the hurricane mass begins to track into higher latitudes and may continue to grow in size and strength. The speed of this tracking movement can vary from nearly 0 to greater than 60 mph. If you consider that the diameter of a hurricane ranges from 60 to 1,000 miles, and that gale-force winds may extend over most of this area, the total energy released by the storm is almost beyond comprehension.

For a hurricane making landfall in the area of Georgia, hurricane forces will be at their maximum in the area to the right (north or east) of the eye, but the entire landfall area will experience severe storm conditions. If the hurricane comes on a high tide, especially a high spring tide, the effects of storm-surge flooding, waves, and overwash will be magnified.

The National Weather Service has adopted the Saffir-Simpson Scale (table 3.4) for rating a hurricane's strength. The scale ranks a storm on three variables: wind velocity, storm surge, and barometric pressure. Although hurricane paths are still unpredictable, the scale quickly communicates the nature of the storm; it gives an idea of what to expect in terms of wind, waves, and flooding.

Don't be misled by such predictions, however. A hurricane is a hurricane. The scale simply defines how bad is bad. In addition, a hurricane can strengthen or change course rapidly and unpredictably. When the word comes to evacuate, *do it*. Don't gamble with your life or the lives of others. (See appendix A for a checklist of things to do when a hurricane threatens.)

Storm Surge

Storm water and wind have well-known effects on beaches and buildings. An additional hazard, sometimes forgotten, is *storm surge*. This

Table 3.3 Monthly Frequency of Tropical Cyclones in Georgia (data from 1898–1968)

	All tropical cyclones	Hurricanes
May	1%	
June	13	
July	9	
August	15	50%
September	36	25
October	24	25
November	2	

Source: Modified from *Georgia Tropical Cyclones and Their Effect on the State*, by H. S. Carter, 1970.

phenomenon is defined as the difference between normal water level and storm-induced water level. Storm surge is produced primarily by high winds and by the low atmospheric pressure of the storm. High winds can push water onshore, while low atmospheric pressure allows water to dome upward, simply because there is less weight from the atmosphere pushing down on the water's surface.

This swell, or dome, of water can cause awesome damage, particularly along a coast with a gently sloping continental shelf, like that of Georgia. During a major storm surge, waves no longer break on the beach; instead, they break inland on dunes or the nearest buildings. In Mississippi in 1969 Hurricane Camille produced an unusual storm surge in which the sea level at the shoreline was as much as 30 feet higher than normal, and the shoreline shifted landward thousands of feet. A storm surge of 17 to 19 feet was estimated for the Great Hurricane of 1898 that pounded the Georgia coast. Even the mainland city of Brunswick suffered 8-foot storm tides (fig. 3.23).

Several other features influence the amplitude and location of maximum storm surge. These include (1) hurricane intensity—the more intense the storm, the higher the surge; (2) hurricane size—maximum storm surge will occur approximately 20 miles north of the point of landfall for categories 1–4 storms (table 3.4); for a category 5 storm, which is more compact, maximum storm surge will occur approximately 12 miles north of the point of landfall; (3) forward speed of the hurricane; (4) bottom configuration (bathymetry) of the area where the surge comes ashore—shallow water enhances storm-surge height; (5) angle of the hurricane's track to the shoreline—storm surge is higher for those hurricanes with tracks perpendicular rather than parallel to the coastline; and (6) physical configuration of the coastline where the surge comes ashore—embayments and estuaries can create a funneling effect, increasing the storm surge's height as it travels upriver or up the bay. Near the head of the estuary, the storm tides may be twice the amplitude of those on the open coasts.

Hurricane Probability

The probability of a hurricane occurring in any one year along any 50-mile segment of the Georgia coast ranges from 1 to 7 percent, with the higher values to the north. The likelihood of a "great hurricane" striking the Georgia coast in any one year is only 1 percent. These low numbers may give a false sense of security, but hurricane history teaches us that in the lifetime of a structure such a storm is almost a certainty. The historical average in Georgia is about one hurricane every 10 to 15 years. Furthermore, the occurrence of a great hurricane in any one year does *not reduce* the likelihood of a similar storm striking again the *next* year.

In addition, recent research indicates that we may be entering a period of greatly increased hurricane activity. Scientists have found a link between climatic conditions in Africa's Subsahel region and the formation of major storms. Dry spells in the Subsahel coincide with times of little Atlantic storm activity,

Table 3.4 Saffir/Simpson Hurricane Scale

Category	Central pressure (inches mercury)	Winds (mph)	Surge (feet)	Damage	Example
1	>28.94	74–95	4–5	minimal	Agnes (1972)
2	28.50–28.93	96–110	6–8	moderate	David (1979)
3	27.89–28.49	111–130	9–12	extensive	Connie (1955)
4	27.18–27.88	131–155	13–18	extreme	Hugo (1989)
5	<27.17	>155	>18	catastrophic	Camille (1969)

Surge amplitudes are specifically calculated for the Georgia coast. These values are higher than those listed in the standard Saffir/Simpson scale because the shallow waters of the Georgia coast emphasize the height of storm surge. Source: Adapted from *Chatham County Hurricane Response Plan*, by Coastal Area Planning and Development Commission (CAPDC), 1982.

3.23 Brunswick, in the aftermath of the "Great Hurricane" of 1898. Photo courtesy of Coastal Georgia Historical Society.

while rainy cycles generally coincide with a higher frequency of large tropical storms. Apparently, these wet/dry spells occur cyclically with a period of about 20 to 30 years. Since we now seem to be entering this cycle's "wet" phase, U.S. coastal homeowners should perhaps brace for more Hugos.

Hurricanes in Georgia

The Georgia coast has not been hit by a truly devastating hurricane since the late 1890s. Naturally, with such a long gap, many coastal residents and developers have lapsed into a state of apathy regarding hurricanes. We shouldn't rest so easy. Meterologists are quick to point to their statistical data and old-timers to their ancestors' tales to remind us that it is only a matter of *when* rather than *whether* a major hurricane will strike the Georgia coast again.

Because of the shoreline's orientation, a relatively small number of intense storms reach Georgia coastal areas compared to more exposed shores (for example, North Carolina). In addition, many of the hurricanes affecting Georgia arrive after passing over the Florida panhandle. This overland journey saps the hurricane's strength, and it usually has diminished to a tropical storm by the time it reaches Georgia.

Georgia, however, has not been totally spared the ravages of major coastal storms (table 3.5; fig. 3.24; appendix B). Since the earliest reported hurricane of 1752, many hurricanes and tropical storms have produced abnormally high winds, intense rainfall, and flooding on the state's barrier islands. With respect to tropical cyclones, Georgia is considered to be at moderately high risk. If the years from 1886 to 1968 can be considered typical, Georgia can expect to receive about one full-fledged hurricane every 10 to 15 years. The effects of tropical storms will of course be felt more often; the average is slightly more than once per year.

Since most current coastal residents have never experienced a major hurricane, we must look to historical accounts to gain some appreciation of the awesome power of these "September gales." In days gone by, our only knowledge of a hurricane's strength came from sketchy newspaper accounts describing the damages and number of lives lost or estimating the losses in dollars. Major storms such as

Table 3.5 Significant Georgia Hurricanes

Date	Classification	Damage
September 1752	Great/Extreme	Extensive damage to low-lying structures and to ships; unknown number drowned.
September 1804	Great	7-foot storm tide in Georgia. Extensive damage to crops and buildings. 3 deaths in Savannah; many more on nearby islands.
August 1813		Severe in Charleston; lesser effect in Georgia.
September 1824	Major	Tremendous freshet in most rivers with extreme crop damage inland. Darien area especially hard-hit.
September 1854	Major	90-mph winds at Savannah. Considerable property damage along Georgia coast. Several deaths.
September 1859		Great destruction and loss of life on Hutchinson Island.
August 1881	Major	Winds 80 mph. 16-foot storm surge. Extensive property damage and 335 deaths in Savannah area.
August 1885	Extreme	Winds 125 mph. Damage apparently light in Georgia.
August 1893	Extreme/Great	Winds 72 mph. 17–20-foot storm tide. Nearly all buildings on Tybee Island damaged. 1,000–2,000 deaths (in South Carolina).
September 1896		Winds 75 mph. Damage heavy on Tybee Island and over much of southeast Georgia. 25 fatalities.
August 1898		Winds 100 mph. Extensive flooding and property damage, especially in Savannah area.
October 1898	Extreme	12-foot storm surge. Damage especially severe in Brunswick/Darien area. At least 180 fatalities.
October 1910	Great	8-foot storm surge. Minor damage. No deaths.
August 1911	Major	Winds 100 mph. Heavy property damage at Tybee Island. No deaths.
September 1928	Great	Beach cottages destroyed.
August 1940	Major	Winds to 90 mph in Savannah; probably higher on Tybee Island. 34 deaths in Georgia and South Carolina.
October 1947	Category II	Winds 100 mph. 12-foot storm surge. 15-foot waves. $1 million damages. 1 fatality.
August 1964	Major	Gale-force winds. 14-foot storm surge. Flooding. 1 death. (Cleo)
September 1964	Category IV	Winds 90 mph. Extensive property damage. No deaths. Federal disaster declared. (Dora)
September 1979	Category IV	Winds 74 mph. Damage light, mostly in Savannah and Liberty County. (David)

Other unlisted hurricanes have affected the Georgia coast as they passed by offshore.
Source: Modified from South Carolina Wildlife and Marine Resources Department, October 1978; House Document 354/86/2; ESSA Technical Memorandum #EDSTM 14; and *Chatham County Hurricane Response Plan*, by CAPDC, 1982.

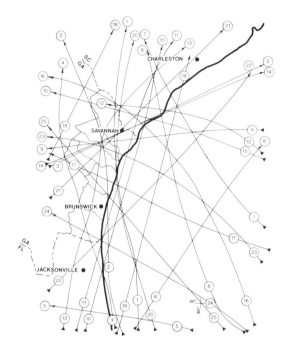

3.24 Tracks of 25 hurricanes that have affected Georgia coastal areas since 1881.

those in 1804, 1854, 1881, 1893, and 1898 were all characterized as "the worst ever" or "greater than" some previous "worst" storm. The hurricanes were not, in fact, increasing in intensity, but each storm followed a somewhat different path: the full fury would be felt by Savannah in one storm and at Brunswick or Darien in the next. During modern times storms have been compared in terms of dollar losses, but this figure describes the damage to structures and facilities rather than the hurricane's strength.

Table 3.5 tabulates recent hurricanes in Georgia. In addition, appendix B briefly recounts some historical accounts of these storms. For more complete tales of storm effects on coastal property and people, see the references listed in appendix D. Local libraries and historical societies also house historical accounts and photographs.

Today, advance warning, efficient evacuation, and safer construction practices *should* result in low casualty rates even in a major hurricane. But unsafe development, allowing population growth to exceed the capacity for safe evacuation, and complacency of coastal residents could reverse this trend with shocking results. Much can be learned from the recent devastation of Hurricane Hugo in South Carolina. Let us emphasize again, the question is not whether such a storm will strike the Georgia coast, but when?

Good News: The Georgia Plan
The 500,000 residents of coastal Georgia are fortunate to be living under the umbrella of the Georgia Plan, a comprehensive and innovative hurricane preparedness program that serves as a model for the rest of the country. The Georgia Plan was begun in 1980 with the help of the Federal Emergency Management Agency and has since been carried on locally by the Coastal Georgia Regional Development Center (CGRDC), formerly known as the Georgia Coastal Area Planning and Development Commission (CAPDC). Unique to the CGRDC program is its coordination of the hurricane evacuation plan not only with the 6 coastal counties which must evacuate, but also with the 23 inland host counties which will receive the fleeing evacuees.

In undertaking its early studies, the CGRDC conducted a telephone survey and found that the people least likely to evacuate were those who had been living on the islands the longest. Recognizing the need to alert residents to the dangers of riding out a storm, the CGRDC launched an effective public information/education program. One of its components is the identification and education of specific target audiences—local government employees and officials, the elderly, schoolchildren, hotel and motel operators, and marina and boat operators. The program's second component involves long-term planning and coordination with local media and the National Weather Service to ensure adequate communication and warning in the event of a life-threatening storm.

The CGRDC also undertook an important mapping project in formulating the Georgia Plan. Two sets of maps were produced: (1) detailed Emergency Operation Center maps for Civil Defense and other emergency managers, and (2) simpler maps for use by the general public. You can find a map outlining evacuation routes, shelter locations, and other pertinent information in your local phone book.

In October 1989 when Hurricane Hugo was approaching the Georgia coast, the Georgia Plan went into effect. A mass evacuation of all islands and of many low-lying areas was ordered. Newspaper reports and radio and television broadcasts gave detailed instructions for storm preparations and evacuation procedures. An estimated 150,000 to 200,000 people evacuated the coast, most from the region near Savannah.

On the evening before Hugo's predicted landfall, the roads from the city were jammed with traffic. Most of the people intending to evacuate left that day. Despite the dire warnings, some people refused to heed them. Within 12 hours of expected landfall, a Civil Defense spokesperson estimated that 15 percent of Wilmington Island's residents had not yet left the island. Fortunately for coastal Georgia, at the last moment the storm veered to the north and into Charleston. The Georgia Plan seemed to work relatively well; however, emergency managers were able to pinpoint weak points in implementing the plan and are working to correct those defects.

For more information about the Georgia Plan, contact the Coastal Georgia Regional Development Center in Brunswick. See appendix C for the complete address.

Stormhounds: A Word of Caution

Nothing can compare with the wild beauty of a winter storm at the coast. There is, however, a fundamental difference between a mild winter storm and a major northeaster or hurricane. Understanding that difference is crucial. There are adventurous (or foolish, depending on your perspective) souls who rush to the islands to experience the excitement of the "big one." One island dweller bragged that he and his eight-year-old son were the last to get *onto* Sea Island just before Hurricane David brushed by in 1979. A Fire Island, New York, resident talked with bravado of the superb vodka gimlet party he hosted as bay waters gradually crept over the floor of his house. Revelers at a hurricane party in Galveston during Hurricane Alicia were forced to leave the Hotel Galvez in the middle of the storm when flying debris threatened their safety. These people were only embarrassed. Others, such as the 35 partygoers (the ones who laughed off a sheriff's warning) in Biloxi, Mississippi, were not so lucky. All were killed.

When word comes to evacuate, *do it*! Don't gamble with your life or the lives of others.

Of *Smilax*, Sabal, and *Spartina*: Patterns of Barrier Island Vegetation

Native vegetation can assist you in reading the history of island areas. Such clues are used in chapter 5 to analyze specific sites for relative safety and suitability for human occupation.

The vegetation you see on a stroll from ocean to marsh across a Georgia barrier island will vary greatly depending on which island and portion of that island you visit. Perhaps the most dramatic differences exist between the "developed" and "undeveloped" coastal islands. Rock revetments and seawalls instead of dunes, housetops instead of maritime forest canopy, and patio-lined swimming pools instead of palmetto-fringed ponds highlight the differences. Yet even on "undeveloped" barriers, different arrangements of plant communities can be noted between an island's north and south ends and between old and young areas of the upland. In general, orderly major zones of vegetation exist on all Georgia barriers.

Let's explore a "generic" Georgia barrier island.

A Walk Across an Island

An ocean-to-marsh cross section of this typical island would reveal the following major zones of vegetation: dunes, transition shrub, maritime forest with freshwater wetlands, and salt marsh (fig. 3.25). Notice that topographical changes accompany the transitions from one zone to the next.

As you leave the shifting sands of the beach, cross the dunes, and proceed toward the forested barrier flat, the relative stability of the sandy island increases. Waves and currents

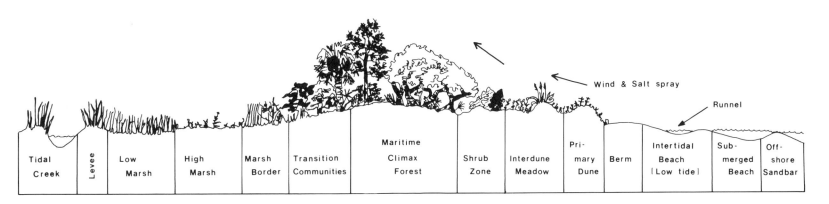

3.25 Typical Georgia island vegetation zonation.

constantly reshape the beach but affect the dune areas only during storms. Farther inland, forested areas are affected only during major storms or through continued long-term erosion. An example of the latter is the bone-yard beach on Wassaw Island (fig. 3.16b).

One important factor that affects where particular plants may live is *salt*. Plants nearest the open ocean must be resistant to salt spray; plants in the marshes must tolerate frequent inundation by saltwater; while those plants unable to tolerate these stresses are confined to the island's interior. (Figure 3.25 and table 3.6 summarize this vegetation zonation imposed by stresses such as sand mobility, salt spray, and flooding.)

The edge of the forest adjacent to the dune system and salt marsh is comprised of communities of transition shrubs such as yaupon holly, wax myrtle, live oak, and *Smilax* (greenbriar). These plants are pruned by the salt spray.

Extensive, well-developed maritime forests distinguish the wilder, less dynamic Georgia islands from other barriers of the U.S. coast. On narrow, highly dynamic barrier islands, extensive maritime forests are unable to develop and mature because of the stresses of rapid shoreline changes and salt spray/inundation. Although characterized by a variety of plant communities, mature maritime forests are often dominated by live oaks and mixed hardwoods, with an understory of palmettos. Forests of pine are usually found on the younger, southern ends and ocean sides of the island forests.

Freshwater wetlands in the forest interior may be seasonally filled sloughs or year-round ponds. Filled by rainwater, these wetlands provide important habitat for alligators, water birds, otters, and other island wildlife. On developed islands, such ponds, if not drained and filled, may be used as water hazards on golf courses.

On the backside of the island are the great expanses of salt marsh. The transition between maritime forest and marsh is abrupt and usually obvious. Plants of the upper marsh can indicate the upper reach of the spring tides.

A walk across most parts of Tybee and Sea Islands takes you past houses and fenced yards instead of through forests. For major portions of St. Simons and Jekyll, the same is true. Destruction of the maritime forests and associated plant communities (as well as freshwater wetlands and topographical features) has accompanied the development of these islands. Unfortunately, little natural area has been preserved on most of the developed islands. Be

Table 3.6 Plant Zonation on a Georgia Barrier Island

Environment	Lower salt marsh	Upper salt marsh	Forest-marsh transition	Freshwater wetlands	Maritime forest	Dune-forest	Backdune	Interdune meadows	Foredune
Plant communities	*Spartina alterniflora*	glasswort sea ox-eye daisy spike grass salt meadow hay *Spartina alterniflora* *Juncus* sp.	sea ox-eye daisy salt meadow hay marsh elder groundsel tree red cedar yaupon holly wax myrtle cabbage palmetto	cattail rush willow	oak red cedar palmetto pine red bay	live oak wax myrtle yaupon holly red bay saw palmetto buckthorn Virginia creeper *Smilax*	wax myrtle saw palmetto buckthorn yaupon holly *Smilax* prickly pear cactus yucca	wax myrtle yucca cabbage palmetto slash pine *Smilax*	sea oats beach grass panic grass beach croton beach pennywort sandspur Russian thistle
Example	Vast expanses between mainland and islands	Along edge of Sea Island causeway	Jekyll Island north end pier	Jekyll Island	Jekyll Island	Jekyll Island south end	East Beach	East Beach	Tybee Island

certain to visit an "undeveloped" island (or at least Jekyll Island) to experience the special feel of a wild barrier island.

Coastal Conflicts: The Collision of Human Plans and Natural Processes

All too often, especially in dynamically active environments, human plans may run afoul of physical processes. For example, clearing the maritime forest does give us a broader, more expansive vista for viewing pleasure, but it also contributes to denuding the island, destroying valuable habitat, and mobilizing loose sand. Building seawalls does diminish the rate of upland erosion, but on a chronically receding shoreline it also spells doom for the sandy beach. The list of examples goes on.

We should now know enough about the natural processes and systems of barrier islands to avoid or at least anticipate and mitigate the effects of such conflicts through wise land use and management. Too often, however, we choose to ignore these guidelines for the sake of personal preference, convenience, and economic gain. In many communities, too, the damage has been done, and residents must cope with a legacy of environmental and economic ills.

We do profoundly alter the environment in which we live. It is unreasonable and unrealistic to aim for "no-impact" communities. However, it is critical to understand the trade-offs we are making whenever we extensively modify our surroundings.

For example, a developer may want to dredge and fill some wetlands to provide boat slips for his condos and perhaps additional upland acreage. His cost-benefit analysis will usually focus on whether his bill for dredging and filling will be greater than his additional profit from the boat slips and additional acreage. What he will not consider—what he has no economic incentive to consider—is how the dredging operations may affect the local fishery in the short term, or how the loss of wetlands may harm the fishery in the long term, or how water quality and flood control may suffer. The developer will not directly pay for these effects, but someone will. It may be the commercial fishermen, or the support services industry for the sports fishery, or the taxpayers when they must step in with clean-up-the-bay programs or expensive flood control projects. But someone will pay.

Ideally, the regulatory agencies that oversee granting permits for such operations will consider these trade-offs and costs. Typically understaffed and underfinanced, however, the regulatory and enforcement agencies are also bound to abide by legislative decree, which may or may not allow consideration of cumulative effects. The developer's one small marina may not profoundly affect stormwater runoff patterns, but what if 10 of his neighbors have the same idea? Can we deny the developer one marina on the grounds that 11 would be too many? The issues of private and public rights and costs are many and complex.

The remainder of this book aims to do three things:

(1) Encourage an awareness and explicit acknowledgment that some trade-offs must be made. While environmental preservationism for its own sake may be an unrealistic goal to pursue on a large scale, we must recognize the extended costs of environmental modification and make informed decisions based on realistic estimates of costs and benefits.

(2) Stimulate consideration of the link between personal actions and community-wide consequences. The right to private ownership also confers considerable responsibility. If we knowingly buy a house on an eroding shore, is it fair to demand that our neighbors (fellow taxpayers) bail us out?

(3) Help homeowners to recognize circumstances where the toll exacted by dynamic processes may be greater than is immediately apparent. Why buy a house on a chronically eroding spit when you could have one nestled high in a maritime forest?

Learning to Live with an Island

Chapter 5, "Selecting a Site on an Island," provides guidelines for identifying and avoiding many of the potential conflicts between human aims and coastal processes. In some cases the existing conflicts on Georgia barriers cannot be changed without major changes in legislation and local priorities. In only a few instances, however, are the problems unavoidable or completely intractable. Developing and implementing strong local and regional land-use management and growth management plans—based on explicitly stated local goals and visions of the desired "flavor" of the community—can aid in preventing new conflicts and easing existing ones.

If you are already faced with the problem of owning a home on an eroding shore, you might ask, "What can I do about my eroding beach?" This question has no simple answer, but it is briefly addressed in chapter 4. The usual bottom line in trying to stop recession of an open-ocean shoreline is that the methods used will ultimately (a) have a negative impact on the immediate or adjacent beaches; or (b) require huge sums of money (usually extensive tax subsidies) to implement and continually upgrade, improve, modify, restore, nourish, or reconstruct; or (c) both. The best response, especially from an environmental and often from an economic standpoint, is to move the threatened structure someplace else.

For a community the retreat from the shoreline at first may not seem economically viable. The long-term prognosis for shoreline changes, however, should govern erosion control and prevention efforts. Certainly, minimal use of engineering structures should be emphasized along with long-term efforts to restore naturally functioning beach systems. Linking regulatory or zoning changes to post-storm reconstruction efforts might be one way to phase in such alterations.

The long-range future of our beaches will depend on how individual communities respond to the rise in sea level and to their migrating shores. Those communities that choose to protect their oceanfront structures at all costs need only look to portions of the New Jersey shore to see what too frequently is the result (fig. 1.4). The life span of houses can unquestionably be extended by "stabilizing the beach" (which really means slowing the upland erosion). The ultimate cost of fighting chronic erosion, however (unless beach replenishment is the only method used), is loss of the beach. The time required for a beach to be destroyed varies greatly and depends on local shoreline or island dynamics.

If when the time comes a community moves the front row of buildings or lets it fall in, the natural beaches can be saved in the long run. Unfortunately, so far in the United States the key factor in shoreline decisions, which every beach community must sooner or later make, has been the availability of funds for erosion control efforts. Communities that cannot generate or obtain these funds allow shoreline recession to roll on. On the other hand, communities that can generate these funds—or ones with pork-barrel clout in Washington—attempt to stop it.

One exception to this rule is Nags Head, North Carolina. Labeled "The Little Town That Could," Nags Head has consciously shaped its planning and growth policies around the results of a poll of the town's property owners. The poll responses made it clear that the townspeople prefer a relatively low-density community with a family atmosphere and a natural shoreline. The result? Beachfront cottages either collapse or are moved inland as their sites fall prey to the area's 2-to-11 feet per year erosion. Yet the town continues to enjoy enviable economic health—with its fine natural beaches intact.

If the past is any indication of where we're going, the future of U.S. shoreline development appears to be one of increasing money expenditures. Most decision-makers now recognize the importance of beaches as storm buffers and recreational resources, and they are opting to pump up artificial beaches instead of building walls. This change in direction is undoubtedly good. However, lawmakers and planners should think even further ahead. The continued commitment of tremendous resources to shoreline stabilization (such as in the Water Resources Development Act of 1986) indicates that little thought is being given to long-term consequences. The future of most developed shores, as presently charted, is one of relatively poor-quality recreational beaches at the price of (1) continual struggles to pump up sand as fast as the ocean takes it away, (2) never-ending expenditures of large sums of public money to protect private development, and (3) hot-and-heavy political battles as coastal communities frantically compete for federal and state bailouts.

Such gloom-and-doom scenarios can be avoided if island management changes to account for probable future trends in beach erosion, sea-level rise, and population growth. By planning future island use instead of merely reacting to current problems, potential conflicts might be eliminated or reduced. Instead of simply living "on" an island, individuals and communities can learn to live "with" it by appreciating the natural processes and changes of an island and a coast.

4 Human Nature and Nature's Shores

Shoreline recession (erosion) is a natural, ongoing process. Although we mark each passing storm with a sudden flurry of "erosion anxiety," we must remember that for the beach the nor'east squall or the rising sea level is no catastrophe. As explained in chapters 2 and 3, for thousands of years Georgia's beaches have survived even the most furious hurricanes—and would continue to do so if allowed to respond naturally. The real enemy is human interference with these natural processes. By constructing our cottages, condominiums, hotels, and shopping centers in the midst of a highly dynamic environment, and expecting them to stay there forever, we ignore the realities of shoreline change. By holding the line, we end up destroying those features that originally drew us to the water's edge.

This phenomenon can easily be observed on the Georgia coast. Along many beaches of the developed islands where communities have chosen to protect the upland and structures near the beach, we find boulders, concrete, steel, and wooden rubble littering their beaches—if there's a beach left at all. In some cases the beaches are so cluttered with the weaponry of defenses against the ocean that the beaches are no longer usable (fig. 4.1). Ironically, to get to the stage of ruined public beaches, often a lot of taxpayers' money had to be spent.

If, however, we travel to an undeveloped place such as Cumberland Island or one of those rare communities where development is in tune with natural processes (such as Little Cumberland Island), we still find beautiful wide beaches (fig. 4.2).

Shoreline Engineering: Trying to Stabilize the Unstable

To better understand how this destruction occurs, let's take a closer look at what happens when we insist that our beaches "Stand still!" *Shoreline engineering* is a general term that refers to methods of changing the natural shoreline system to stabilize it (that is, hold it in place). These methods can range from simply planting dune grass (fig. 4.3) to installing complex, expensive, massive seawalls or groins (fig. 4.4).

4.1 The hazards of shoreline engineering, Tybee Island, 1985. This sign at the north end of Tybee's multimillion-dollar artificial beach project warns swimmers of the dangers created by erosion control structures. Photo by T. Clayton.

4.2 The natural beaches of Little Tybee Island: swimmer's paradise! Tybee Island can be seen in the background. 1976. Photo by L. Taylor.

Seawalls and Bulkheads

Seawalls and *bulkheads* are structures built parallel to the shoreline. Their purpose, which is often misunderstood, is *not* to protect the beaches in front of them, but rather the buildings or upland property behind them. If your community has opted to put in seawalls on a chronically eroding shore, it has, in effect, decided to sacrifice the natural beach.

Seawalls (fig. 4.4) are massive structures, designed and built to withstand the ocean's full impact during each tidal cycle. In short, seawalls are designed to hold back the sea. Bulkheads, which are in effect small (and sometimes relatively flimsy) seawalls, are designed and built to hold back the land and to withstand the force of occasional storm waves. On the Georgia coast bulkheads are often installed along riverbanks and marsh shorelines.

Well-designed seawalls are, in fact, quite effective in fulfilling their intended purpose: saving buildings. Many buildings along the U.S. shore owe their existence today to a protective seawall. However, erecting either a seawall or a bulkhead on a retreating open-ocean beach is a drastic measure, for there are predictable and undesirable side effects associated with their use.

(1) Seawalls reflect wave energy. As the surf

4.3 Tybee Island Dune Restoration Project, 1989. Photo by T. Rust.
4.4 Seawall and groins at the south end of Tybee Island, 1989. Photo by T. Rust.

4.5 Sea Island sloping seawall, 1981. A sloping seawall may diminish, but will not eliminate, the negative impacts of wave reflection. Note the absence of a high-tide beach here.

breaks against the vertical structure, wave energy is deflected and reflected downward and seaward, which may actually accelerate beach erosion. A sloping seawall (fig. 4.5) may diminish this effect to some degree, but it will not eliminate it. Beach removal and scour in front of the seawall may result in the beach's offshore profile being steepened, or at least deepened, which allows increasingly larger waves to strike the wall with direct force. Larger waves in turn increase erosion, which further steepens the offshore profile. And the cycle continues.

(2) Seawalls may increase the intensity of longshore (surf zone) currents, causing the beach to erode faster.

(3) Seawalls prevent the free exchange of sand within the beach/dune system. When seawalls are in place, the beach cannot as easily supply new sand to the island's dunes, nor can the dunes provide sand for the protective beach and offshore bars during storms as would happen within a natural system.

(4) Seawalls create an end-around effect, concentrating wave and current energy at the ends of the wall, thus increasing erosion at these points.

(5) Seawalls are a safety hazard to swimmers, surfers, and other beachgoers (fig. 4.1). One former sea kayaker we met in Florida told us he gave up the sport altogether after getting caught in a sudden squall with nowhere to come ashore. The natural beach had been completely replaced by seawalls, against which he and his boat were dashed by the waves.

(6) Seawalls cannot respond to the presently rising sea level.

(7) Protection of upland and structures by a seawall may actually stimulate an increase in development or redevelopment behind the "safety" of the wall. Often the hazards that prompted construction of the wall are completely forgotten.

(8) Most important, perhaps, is the fact that installation of the first seawall is usually an unspoken but de facto long-term commitment to a never-ending series of increasingly bigger ("better") stabilization projects. As the beach in front of it disappears, a seawall on an eroding shore will deteriorate and must eventually be replaced by one even more massive. The only alternative to repetitive destruction and reconstruction of the seawall is the repetitive placement of sand seaward of the wall to protect it. This option is gaining increasing attention in Georgia and elsewhere.

The state that bears the dubious honor of having led the field in shoreline stabilization has "loaned" its name to this process of ever bigger and "better," which results in extreme "hard engineering" of the shoreline. This is how an official of Monmouth Beach, New Jersey, describes the *New Jerseyization* of the town's beach: "There were once houses and even farms in front of that wall. First we built small seawalls and they were destroyed by the storms that seemed to get bigger and bigger. Now we have come to this huge wall which we hope will hold." The huge wall of which he spoke is high enough to prevent even a glimpse of the sea beyond. A climb to the top reveals no beach, only waves lapping at the foot of the wall (or crashing over the top, depending on the weather) and the remnants of former seawalls, groins, and bulkheads for hundreds of yards to sea.

The good news is that the national trend in

coastal management and engineering is away from such destructive measures and toward gentler means of minimizing beachfront property losses. Three states, Maine, North Carolina, and South Carolina, have taken the important step of banning all such hard structures on their beaches, thereby formally committing themselves to a program of natural beach preservation.

Riprap and Revetments

If you take an aerial flight along the Georgia coast, you will undoubtedly be struck by the fortresslike appearance of some of the islands (fig. 4.6). Rimmed with a mantle of boulders, Jekyll, St. Simons, and Sea Islands have indeed declared war on the sea.

The rock that you would see is called *riprap* or *revetment*, and islanders use it to protect shorefront upland property from erosion. Both terms refer to large blocks of rock placed on or at the back of the beach to absorb and reflect wave energy. Riprap is simply rock stacked directly on the beach. A revetment is designed and constructed with more care and deliberation. It is first underlain by a porous basal filter cloth and a layer of small rock before large, armoring boulders are stacked on top. These large, loose rocks are intended to reflect and absorb wave energy, protecting the property behind them from direct wave attack. On Tybee and Sea Islands, riprap is placed at the

4.6 Erosion control fortifications on St. Simons Island. Photo by W. Cleary.

seawall toe in some places to prevent scour or undermining and potential collapse of the structure (fig. 4.7).

Emplacing riprap or a revetment, like erecting a seawall or bulkhead, is an extreme measure. You should know what to expect when you install riprap or a revetment on an eroding shore, or if you buy property already equipped with boulders on the beach.

(1) Revetments and riprap, like seawalls, reflect wave energy. The principal difference between the loose rock and a seawall is that the open structure of the riprap (or revetment), in contrast to an impermeable seawall, can absorb more wave energy. This partial absorption is important in reducing scour at the toe of the structure. The filter cloth and graded structure of a revetment lends even greater protection from undermining and settlement or collapse of the structure. The bottom line, however, is that there will still be some wave reflection off the rock, especially during storms, when absorption may be relatively unimportant.

(2) Revetments may increase the intensity of longshore currents. Again, like seawalls, revetments restrict longshore currents and may enhance their scouring effects.

(3) Revetments, like seawalls, prevent the

free exchange of sand within the beach/dune system. Without this crucial flexibility, the beach on a receding shore is doomed to eventual death.

(4) Revetments create an end-around effect. Like seawalls, the rock rubble changes incoming wave patterns in such a way that erosion actually increases at the ends of the structure. An example is the Goodyear house on Sea Island. Here, one individual singlehandedly resisted an effort to construct a community seawall and refused to build a wall or revetment on his property. However, the end-around effect of adjacent seawalls speeded up erosion at the wall's ends, threatening his home with collapse. The next owner received permission from the state to build a wall and filled in the notch.

4.7 Riprap is sometimes placed at the toe of a seawall to prevent scouring and undermining of the structure. Sea Island, 1985. Photo by T. Clayton.

(5) Revetments are a safety hazard. Common sense tells us that placing big walls of huge rocks in places where people swim, stroll, and play is dangerous. This point was tragically illustrated a few years ago at an exclusive South Carolina resort when a young mother was dashed against a riprap wall and killed. Aside from the difficulty of simply getting over and around a huge pile of rocks to get onto the sandy beach, there is the additional problem of properly sizing the individual rock fragments. If the pieces of rock are too small to withstand the impact of ocean waves, the result is a beach strewn with rock rubble. The residents of Sea Island learned this lesson through experience; portions of the naturally sandy Tybee and Jekyll beaches have been rock-strewn at times.

Being topped by storm waves is a potential problem with any beach structure, but it becomes a particular hazard if the overtopping waves hurl bits and pieces of your revetment.

(6) Revetments cannot respond to the presently rising sea level.

(7) Finally, there is the ever-present question of the irreversibility of hard stabilization. As with seawalls, you must give considerable thought to long-term prospects. Is the presence of a natural recreational beach important to you? If so, you should avoid any such hard stabilization and consider other options available to minimize property loss. And what will you do when the initial structure is damaged? Are you prepared for a future of continual maintenance and expense?

Chances are that most beachfront property owners will be strongly in favor of protecting their houses and properties by some means. When looking at it from a broader viewpoint, however, the question should be asked: "Should the quality or even the existence of the public's natural recreational beach be compromised or sacrificed for the sake of erosion protection for the privately owned upland?"

Groins

Groins, which are walls constructed perpendicular to the shoreline, are designed to build out a protective beach by trapping a portion of the sand being transported by longshore currents (fig. 4.8). *Holding groins* are a variation of *trapping groins*. These groins are not intended to trap the natural littoral drift, but to hold an artificially placed beach where it is desired. The north terminal groin and the newer south terminal groin on Tybee Island are examples of such holding groins (fig. 4.9). With time, holding groins are apt to become trapping groins.

The history of groin construction along the Georgia coast has been lively, with more than 120 built in the past century on Tybee alone. Most recently, three removable groins have been built in front of the Beach Club property on Sea Island. These groins are removable so that the holding groins do not become trapping groins.

Although most groins are now constructed

of rock or cement, many materials have been used in the past. On Tybee Island in the early 1900s bundles of brush (trees and shrubs) often served the purpose. Later, timber was a favored material. All such groins were short-lived.

If a groin works as intended, sand will pile up and produce a high and wide beach on the updrift side. However, the sand trapped and held by that successful groin was probably

4.8 Groins are designed to trap and hold sand that is being transported by longshore currents.

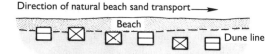

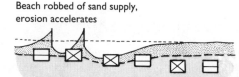

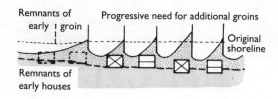

4.9 Terminal groin built to hold in place Tybee Island's artificial beach. View from Tybee Island lighthouse, 1989. Photo by T. Rust.

flowing downdrift to a neighboring beach. Deprived of this alongshore supply, the downdrift beach will erode faster. Erosion downcoast from a groin or groin field is common and should be expected (fig. 4.8). Lawsuits brought by angry homeowners for this reason are not unheard of. Be forewarned that the courts have held builders of groins liable for the damage they cause.

On Tybee Island irate property owners accused the Corps of Engineers of exacerbating beach erosion by its placement of the south-end groin. These individuals own property south of that groin—outside the limits of the federal beach replenishment project—and they blame erosion that has occurred since the groin was built on sand starvation.

Rip currents have been associated with groins, and if the groin is unusually long, sand may be shunted far enough seaward that it is permanently lost from the beach system. These rip currents are also a safety hazard to swimmers.

Although under certain conditions groins can temporarily form or stabilize a beach, they cannot in themselves completely protect against wave action. Under the "right" conditions, waves can remove sand from between the groins and carry it directly offshore, leaving property exposed and unprotected.

Groins also fail through detachment. If land at the groin's shoreward end is eroded, sand can bypass the groin and render it useless. Such detachment usually occurs during storms when the groin is needed most.

Finally, groins, like other forms of shoreline stabilization, cannot respond to the rising sea level.

Jetties

Like groins, *jetties* are built approximately perpendicular to the shoreline. Their purpose, however, is different. Groins are built to control beach erosion on their updrift side, jetties to stabilize navigation channels by preventing sand from flowing into them at the inlets. To stymie such transport of littoral sediments, jetties must be very long, extending beyond the surf zone. In effect, they act as huge groins, trapping sand and diverting it far offshore. Jetties thus profoundly influence an area's erosion and accretion patterns (fig. 3.19).

In fact, most erosion trouble spots along Florida's east coast are created by jetties. The St. Marys' jetties at the Georgia-Florida border are among the largest on the U.S. East Coast. They have caused massive amounts of sand (about 150 million cubic yards) to be transported offshore—sand that belonged in the nearshore beach sand system. Because sand here flows largely from north to south, the detrimental effects of the St. Marys' jetties are felt almost exclusively in Florida—a stroke of luck for Georgia's Cumberland Island and a severe problem for Florida's Amelia Island dwellers.

Breakwaters

Although there are no breakwaters along the Georgia coast, these protective structures are likely to be proposed among the solutions to shoreline erosion. Breakwaters intercept waves and reduce their energy before they reach the shore. Reduced wave energy allows sand to be trapped in the "shadow" of the breakwater, building a protective beach (fig. 4.10). The obvious benefit is reduced erosion and beach growth behind the breakwater.

This protection, however, comes at the expense of neighbors. By interrupting the wave pattern, the breakwater reduces the longshore current's strength, allowing sand deposition behind the structure. Moreover, this buildup acts as a barrier to additional sand transported by longshore currents and leads to narrower beaches and greater erosion along the downdrift shore (fig. 4.10).

By artificially bypassing sand to the downdrift shore this effect can be reduced, but such systems add to the breakwater's design and operating costs. In addition, the break-water itself may cause scour on the structure's seaward side leading to its failure. Sometimes wave modification by the breakwater may in-tensify wave energy on nearby beaches, adding to the erosion problem. Safety is another consideration. For example, a breakwater in Pinellas County on Florida's west coast has been the site of an unusual number of drownings and near-drownings and is cur-rently embroiled in controversy.

4.10 Overhead view of a breakwater.

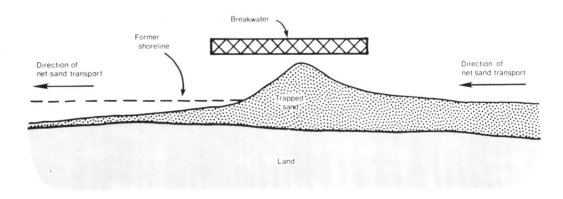

Beach Replenishment

Beach replenishment, which is being embraced by some as the final solution to America's shoreline problems, consists of pumping or dumping sand onto a beach to form a wide and high berm, and sometimes a dune as well (fig. 4.11). Tybee Island and Sea Island have artificial beaches, and nourishment projects are being considered for St. Simons and Jekyll Islands.

A new artificial beach offers considerable protection, at least temporarily, to beachfront property from future storms; it also creates a recreational beach for the public (figs. 4.12a, 4.12b). Beach replenishment is definitely the preferred means of shoreline stabilization, but it still has its problems.

The first step in replenishment is finding a source of suitable sand to pump onto the beach. This task is not always easy. Some communities on the Outer Banks of North Carolina experience the highest erosion rates on the East Coast; but because no affordable sand supply is available, they have resorted to alternate methods of dealing with shoreline retreat (such as moving houses back).

It is important to keep in mind the distinction between methods that truly introduce "new" sediment into the littoral drift system and those that merely redistribute what sand is already in the system. Tidal shoals, for example, are as integral a part of the littoral drift system as beaches. Their configuration reflects some state of dynamic equilibrium that is maintained by waves and currents. If a dredging operation alters a shoal system's shape or volume, it is reasonable to expect that those waves and currents will act to return the system to its state of equilibrium.

Among the possible sources of sand in the coastal zone, most have potentially serious drawbacks. A common and relatively inexpensive source of replenishment material used to be the estuaries on the backsides of barrier islands. Dredging to obtain the sand, however, adversely affects estuarine plants and animals and is currently seldom allowed. Another problem is that sediment from the bay is usually too fine for use on an open-ocean beach; it is often muddy and quickly eroded away.

Another option is excavation from the mainland, as at Myrtle Beach, or from a pit on the island itself, as at Hunting Island, South Carolina. However, excavation from the island contributes to the overall loss of island volume by taking precious sand from one location and placing it on the beach to be eroded away. Sand from a mainland source may be a bit more expendable, but the transportation costs can be prohibitively high, especially with rising energy prices.

A sand source that is attracting more and

4.11 Beach replenishment.

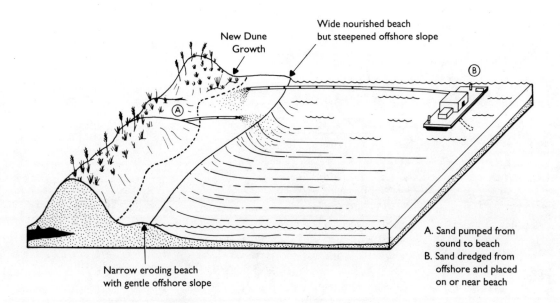

A. Sand pumped from sound to beach
B. Sand dredged from offshore and placed on or near beach

48   Living with the Georgia Shore

4.12 North-central end of Tybee Island, Highway 80/Butler Avenue area. (a) After the 1975–76 replenishment, but before the 1987 replenishment. (b) A few months after the 1987 replenishment. February 1988. Photo by T. Clayton.

more attention is the continental shelf. With relatively little (or at least not obvious) environmental harm, dredging offshore can yield sand that resembles natural beach sand. However, this method is also the most expensive because heavier equipment is required to withstand open-ocean waves and because the sand often must be pumped from great distances offshore.

Also, digging a hole on the shelf may affect wave patterns on the adjacent shoreline and increase erosion rates. Off the Connecticut coast, for example, wave patterns were altered by a hole dredged on the shelf, and as a result the precious replenished beach rapidly disappeared. The same has happened at Grand Isle, Louisiana, and perhaps at Hollywood Beach, Florida.

Dredging offshore also creates pits that may refill with mud and silt (as at Jupiter Island, Florida, and off the south shore of New York's Long Island). If a storm disturbs the fine sediment accumulated in these pits, water turbidity problems result. (Muddier-than-normal water adversely affects some marine organisms.) This phenomenon has been observed off Long Island and may also be occurring at Delray Beach in Florida as a result of beach replenishment programs.

Inlet shoals are still another source of sand for replenishment and have been used at Tybee and Sea Islands. However, dredging island spits or inlet shoals may alter wave or current patterns and result in unpredictable patterns of erosion. In 1976 the Corps of Engineers dredged Tybee Creek shoals, 2,000 feet off the southern end of Tybee Island, to provide sand for the island's multimillion-dollar replenishment project. Within six months the island's southern tip began to experience unprecedented and unpredicted high rates of erosion. An investigative study drew no links between the dredging and the erosion. A field study of Tybee Island's southern shoal system indicated that sand was transported seaward away from the beach/shoal system by ebbing currents that flow through the shoals. Cyclical movement of the channels between the shoals was cited as contributing to the erosion and accretion patterns on the island's south end. This study, however, was conducted in 1978 (two years after the episode of high erosion) during a different season of the year and over a short period. Therefore, it may not have adequately captured local conditions or processes at the time of the rapid erosion.

Some scientists infer that the dredging and erosion may be related because of the temporal link between the dredging of the shoals and erosion of the beach. What the islanders call "just common sense" suggests that the dredged hole acted as a sink for the beach sand as it quickly filled to return the shoal system to its natural state.

Besides inlet shoal systems, a few other sand sources for beach replenishment are available along the Georgia coast. These include river shoals, sands from adjacent beaches and islands, and sediment from maintenance dredging of navigation channels and harbors. All of these sources have been considered for the Tybee project and for the proposed Glynn County projects.

As mentioned, care must be taken to avoid creating or exacerbating erosion problems when obtaining sand from nearshore or offshore shoal systems. For instance, incorrect dredging at Pelican Shoals—a possible source for Sea Island nourishment—might prompt erosion problems elsewhere. (Certainly, the thousands of seabirds that use the shoal site will be disturbed.) In recent years inlet shoals were rejected as a sand source for the Ocean City, Maryland, replenishment project because of potentially adverse affects on adjacent shores.

River shoals may provide sand for the St. Simons beaches. Another potentially important source is navigation-related dredged sediments that could be reintroduced into the littoral drift system. Florida has taken steps to prevent offshore dumping of any more precious littoral sediments. Mechanically bypassing sand trapped by channel jetties is another potential source for downdrift, sediment-starved beaches.

Sand quality also can be a problem for replenished beaches. It is not easy to find a sand source that exactly matches the natural beach sand, so the artificial beach will likely have sand that is finer or coarser. Another common feature of replenished beaches are large clayballs, which are sometimes present on Tybee's artificial beach. If the sand is too fine, several problems can occur. The least serious is sand blowing ashore, as at Delray Beach. There, swimming pools were half-filled with sand, and lawns and even Highway A1A were sometimes covered. Depending on your perspective, however, the nuisance of windblown sand may be a small price for the protection the new beach affords.

The more serious consequence of using sand that is too fine is the possibility that the new beach will be rapidly lost. The National Park Service at Cape Hatteras was becoming familiar with this unhappy experience before it established a policy in 1973 of allowing the barrier islands to respond naturally to coastal forces. One Cape replenishment project which used sand that was too fine disappeared before the contractor had even finished pumping in sand!

It is important to understand that the proper grain size of sand is vital. But it is not the only important element. The next stormy season may take out a replenished beach, no matter how carefully the grain size and other design specifications have been engineered. Ocean City, New Jersey, for example, pumped up a new beach in 1982. Unfortunately, the replenishment coincided with a stormy year (storms are not unexpected on the coast, remember), and in just 2½ months the $5.2 million beach was no more! As one local geologist explained, "Ocean City could have had golf balls on that beach, and they would have been gone."

Sand that is too coarse can detract from the beach's recreational quality. Bathers at Presque Isle, Pennsylvania, for example, complain that recent nourishments there have resulted in beaches no longer suitable for swimming. This is unlikely to be a problem in Georgia.

Like seawalls, an artificially wide beach may offer an unwarranted illusion of "safety" and spur intensified development. If you're thinking of buying a house behind an artificial beach, be sure to find out how long ago the sand was pumped up and how long it has lasted after previous pumping operations. Some areas of some artificial beaches are relatively stable, while others may quickly erode. You may see lots with chronic erosion problems put up for sale immediately after a renourishment operation, and they will look reasonably safe, far from the ocean's reach.

But that will prove an illusion. Remember, too, that funding for future renourishment is not preordained. Even congressional authorization of a project merely grants permission; funding is not guaranteed.

The bottom line in beach replenishment (and other forms of shore stabilization) is the universal determinant—money. Unfortunately, beach replenishment is extremely expensive and, even worse, unpredictably so. Even allowing for inflation, replenishing Carolina Beach, North Carolina, will cost taxpayers 7 to 12 times more than predicted by the Corps of Engineers when Congress originally authorized the project in 1962.

Costs are unpredictable because coastal processes are unpredictable. So far, coastal engineers and scientists do not understand nearshore events well enough to forecast how long a replenished beach will last.

Accordingly, long-term estimates of volume requirements are untrustworthy. Congress in 1962 authorized a replenishment project for Wrightsville Beach, North Carolina, with the understanding that 31,800 cubic yards of sand every year would be needed to maintain adequate hurricane protection. However, unforeseen consequences of natural processes and man-made modifications (closing one inlet and jettying another) have resulted in taxpayers being charged to pump up enough sand to provide (at 31,800 cubic yards per year) what should have been more than 120 years' worth of beach!

The Tybee Island beach erosion control project, like most others, has a 50-year "life span." The initial project design called for renourishment on an average of once every three years. Since the initial replenishment in 1975–1976, only one renourishment has occurred (in 1986–1987). Total cost: approximately $13 million. At first glance, this project might appear to be performing even better than expected—with only one renourishment after 15 years. In reality, however, the beach needed to be renourished years before the operation was actually undertaken. The holdup came down to the availability of funds. In Tybee's case, as well as in others, federal and state funds can be difficult to obtain. That is, the bottom line is often the bottom of the barrel.

Whose Cost? Whose Benefit?

Rarely is a stabilization project proposed without a call for public funding, especially if the project is large. Since responsible citizenship requires critical review of public expenditures, these questions must be investigated and answered: (1) Whose cost? (2) Whose benefit?

Generally, the cost is borne by the public. Is our society willing to continue spending large (unpredictably large) sums of money to (a) build seawalls and revetments that often destroy the recreational and protective beach? and (b) pump up sand again and again forever and ever? Who benefits from the project? Property owners only? Beachgoers too? How long do these benefits last? Are these benefits safeguarded? For example, is there adequate public access to a publicly funded artificial recreational beach? Will that access be maintained?

Typically, the federal government pays the largest portion of the bill, with a smaller share paid by the state, and still smaller shares paid by the county, municipality, or community receiving the sand or the seawall. Carolina Beach, for example, received a new $10.4 million beach in 1982. The town's share? $474,216. "I think it's a bargain," said the town manager.

Cost-sharing for the Tybee project has been 51 percent federal and 49 percent state. Minor amounts were initially contributed by Chatham County and the City of Tybee.

The benefit to the general public can be highly variable, depending on which erosion control method is used. Often, projects have seemed to benefit only the private property and beachfront homeowners who are experiencing upland erosion. In contrast to some other types of projects beach replenishment does offer obvious advantages in that it attempts to provide at least some beach for the recreational user.

A Modest Proposal

Construction too close to the shoreline means great expense to government agencies and a tremendous outflow of tax dollars. Not only are there the direct costs of building roads, bridges, sewer systems, and other services (and often rebuilding them after each storm), but there are huge indirect costs associated with regulating, permitting, monitoring, and enforcing. In addition, regulations imposed by governments in an effort to protect the public interest and cut public losses are often resented and vigorously opposed by those subject to the regulations.

How about a fresh approach altogether? An alternative to the traditional concepts of "regulate and enforce" might be to allow essentially unregulated development at the beach (unregulated in the traditional sense), *provided that*:

(1) Developers, owners, residents, politicians, and community officials may do *nothing* that will harm the public's beach. No seawalls, no groins, no debris, no sand of lower quality than the natural sand would be permitted.

(2) Some degradation of water quality and other environmental destruction inevitably accompany even the most cautious development. Any resource degradation or consumption must be *paid for* by those responsible. (Many Americans are not used to thinking in these terms in regard to environmental problems. For a good introduction to these concepts, read *Environmental Improvement Through Economic Incentives*, which is fully cited in appendix D.)

(3) No public expenditures for development in an environment obviously unsuited for development would be permitted. No government subsidies for infrastructure, shoreline stabilization (engineering), insurance, or disaster relief would be allowed. In other words, provisions of the federal Coastal Barrier Resources Act (discussed in chapter 6) would be extended to include all barriers, not just those declared "undeveloped" in 1982.

At first glance, some of these suggestions may seem outrageous. All of them, however, have been adopted or are being considered in one form or another for application to specific issues in limited areas. In any event, careful consideration and vigorous discussion along these lines may prompt other innovative approaches. In the meantime, concerned individuals must continue to take timely action along conventional paths to encourage prudent coastal management.

Questions to Ask If Shoreline Engineering Is Proposed

When a community is considering some form of shoreline engineering (table 4.1), it almost invariably does so in an atmosphere of crisis. Buildings and commercial interests are threatened, time is short, an expert is brought in, and a solution is proposed. Under such circumstances the right questions are sometimes not asked. You might raise the following questions if you find yourself a member of such a community.

1. Will the proposed response to shoreline erosion damage the recreational beach? in 10 years? in 20 years? in 30 years? in 50 years?

2. How much will it cost to maintain the proposed project for 10 years? for 20 years? for 30 years? for 50 years? Where will the funds come from?

3. If the proposed response is carried out, what is likely to happen in the next big storm? mild hurricane? severe hurricane?

4. What has this shoreline's erosion rate been during the past 10 years? 20 years? since the late 1930s (the time of the first coastal aerial photography)? since the mid-1800s (the time when the first accurately surveyed shoreline maps were produced by the old U.S. Geodetic Survey)?

5. What will the proposed response do to the beachfront? Will the solution for one portion of a shoreline create problems for another portion? How will adjacent islands be affected?

6. What will happen if an adjacent inlet migrates? closes up? What will happen if the tidal delta offshore from the adjacent inlet changes its size and shape or if the channel moves?

7. If the proposal is carried out, how will it affect the density and nature of future beachfront development or redevelopment? Will additional controls on beachfront development be appropriate as part of the solution?

8. What will happen 20 years from now if a nearby inlet is dredged for navigation? if jetties are constructed? if seawalls and groins are built?

Table 4.1 Methods Used to Reduce Erosion and Erosion Loss

Move building	Beach nourishment	Landscape shore	Seawalls[1]	Groins	Breakwaters
ADVANTAGES					
Reduces threat of loss	Reduces erosion by increasing beach width	Increases stability of slope	Reduce erosion by armoring shore	Reduce erosion by trapping sand and increasing beach width updrift of structures	Reduce erosion by reducing wave energy reaching shore
Allows natural shore processes	Allows natural shore processes	Allows natural shore processes to operate			Increase beach width behind structures
May be less expensive than shore structure	May increase attractiveness of shore	May increase attractiveness of shore			
	Increases beach width downdrift				
DISADVANTAGES					
Does not reduce erosion	Periodic replenishment will be necessary	Does not materially reduce erosion	Destroys beach on chronically eroding shore	Impede the longshore transport of sand, causing erosion downdrift	Impede the longshore transport of sand, causing erosion downdrift
Sufficient area must be available for move	Expensive	Sufficient area must be available for landscaping	May limit access and recreational use of beach; safety hazard	May limit beach use; safety hazard	Need periodic maintenance
Building must be movable	May negatively affect ecosystem	Expense variable	Need periodic maintenance	Need periodic maintenance	More expensive than seawalls or groins
			May reduce attractiveness of shore	Need abundant sand in longshore system	May reduce attractiveness of shore
			Disrupts dune/beach sand-sharing system	May reduce attractiveness of shore	Safety hazard

[1] Includes bulkheads, revetments, and riprap.

9. What is the 50- to 100-year environmental and economic prognosis for the proposed response if predictions of an accelerating sea-level rise are accurate?

10. If stabilization—for instance, a seawall—is permitted, will this open the door to seawalls elsewhere? (The answer to this one in most other coastal states has always been "yes".)

11. Are alternatives to engineered solutions available? What are the pros and cons of allowing the beach to remain in or return to a natural state?

Truths of the Shoreline

From studying beaches around the country, we can see that certain generalizations or "universal truths" clearly emerge. These are equally evident to scientists who have studied the shoreline and to old-timers who have lived on the coast all their lives. Unfortunately, these principles are not as clear to the family vacationing for the summer or to the couple condo-shopping for the weekend, and these people are often victims of their own innocence. "How were we to know?" is a homeowner's lament commonly aired on TV news stories about beach erosion. As aids to safe and aesthetically pleasing shoreline development, these truths should serve as fundamental guides for planning on any barrier island.

1. *There is no erosion problem until a structure is built on a retreating shoreline.* Beach erosion is a common, expected event, not a natural disaster. Shoreline recession is generally not a threat to barrier islands. It is, in fact, an integral part of island evolution (see chapter 3) and of the barrier's dynamic system. When the beach retreats, it is not disappearing but simply moving landward. (There are rare exceptions to this rule, where local sea level is rising very rapidly in an area of extreme sediment starvation—as in Louisiana.) Whether the beach is moving landward or seaward does not concern the visiting swimmer, surfer, hiker, or fisherman. It is only when people build "permanent" structures in this zone of change that problems develop.

2. *Construction too close to the shoreline causes shoreline changes.* The beach exists in a sensitive balance with sand supply, wave energy, and sea level. (This dynamic equilibrium is discussed in chapter 3.) Construction on or near the shoreline can change this balance and reduce the beach's natural flexibility. The result is change that often threatens man-made structures. Dune removal, for example, reduces the sand supply used by the beach to adjust its profile during storms. Revetments and seawalls essentially eliminate this sand exchange between beach and dunes altogether. Jetties drastically alter the (longshore) movement of sand. Debris from collapsed cottages and failed engineering structures also contributes to property damage and may even increase shoreline erosion during storms.

3. *Shoreline engineering protects the interests of a few, often at a high cost in federal and state dollars.* Shoreline engineering is generally carried out to save beachfront property, not the beach itself. Beach stabilization projects are undertaken in the interest of the minority of beach property owners rather than the general public. If the shoreline were allowed to migrate naturally, the fisherman and swimmer would not suffer. Yet beach property owners apply pressure for spending tax money—public funds—to protect the upland with walls or artificial beaches. Do their personal interests alone warrant the large public expenditures required for shoreline stabilization?

Beaches near large metropolitan areas may be exceptional cases. The combination of extensive high-rise development and heavy beach use (100,000 or more people per day) may economically justify extensive and continuous beach replenishment projects.

Even the existence of expensive high-rises, however, does not necessarily justify spending such large public sums. In Miami Beach, for example, the owners of high-rise hotels initially opposed replenishment because federal involvement meant greater public beach access. Eventually, however, the $54 million beach was built. This publicly funded beach has now been responsible for a real estate boom and tremendous prosperity for the property owners of Miami Beach. One cannot help but wonder why the truckdriver in Oklahoma is contributing tax money to the economic well-being of Miami Beach luxury hotels.

Again, why not have those who reap the benefits pay the costs?

4. *On a receding shore, "hard" stabilization destroys the beach.* If this sounds incredible to you, drive to New Jersey and examine that state's shores (fig. 1.4). See the miles of "well-protected" shoreline—without beaches! Or drive to Tybee, Sea, St. Simons, or Jekyll Island.

5. *The cost of saving beach property through shoreline engineering is sometimes greater than the value of the property to be saved.* For a variety of reasons price estimates in the long run are often unrealistically low. Maintenance, repairs, and replacement costs are typically underestimated because it is erroneously assumed that the big storm, capable of removing an entire beach replenishment project overnight, will somehow bypass the area. The increased potential for damage resulting from shoreline engineering is also ignored in most cost evaluations. Shoreline engineering projects would be funded with somewhat greater reluctance if those controlling the purse strings realized that such "lines of defense" must be perpetual.

6. *"Once you begin shoreline engineering, you can't stop it!"* This statement, made by a city manager of a Long Island Sound community, is confirmed by shoreline history throughout the world. Once a "hard" shoreline structure is installed and fails, "better"—that is, larger and more expensive—structures must later be installed, only to suffer the same fate as their predecessors. Or eventually the protective structure itself must be protected by an artificial beach. Even with "soft" engineering approaches, people develop a sense of perpetual entitlement to aid.

History shows us two situations that may terminate shoreline engineering. First, a civilization may fail and no longer build and repair its structures. This was the case with the Romans, who built mighty seawalls. Second, a large storm may destroy a shoreline stabilization system so thoroughly that people decide to walk away. In the United States, however, such a storm is usually regarded as an engineering challenge. Accompanied by cheers of encouragement and admiration for the spirit of those picking up the pieces, and undaunted by the lessons of the sea, government agencies and individuals rush to rebuild those structures just destroyed, even those unwisely sited. This flurry of post-storm reconstruction is often largely funded by emergency or disaster relief funds from state and federal agencies and by federally subsidized insurance payments. And so the cycle goes on. . . .

Some Guiding Principles

1. Design to live with the flexible island environment. Don't fight nature with a "line of defense."

2. Consider all man-made structures near the shoreline temporary.

3. Accept as a last resort any engineering scheme for "beach stabilization," and then only for metropolitan areas. Reject "hard" stabilization.

4. Base decisions affecting island development on the welfare of the general public rather than the minority of shorefront property owners.

5. Let the lighthouse, beach cottage, motel, or hot dog stand fall when its time comes. Or move it inland.

6. Save our natural beaches for the next generation.

5 Selecting a Site on Georgia's Barrier Islands

An important principle of geology is often expressed as "The present is the key to the past." Whether we're looking back to a time millions of years ago, trying to understand how a magnificent mountain came to be, or whether our perspective is a mere hundred years as we try to comprehend how a grassy dune field formed, the processes we observe today can help unravel the geological story of those years gone by.

Conversely, the past can provide us with clues to the present and even the future. For example, coastal scientists have noted that coastal areas severely damaged in the last storm are often the same areas hit hard in the storms before; these same areas are likely to be adversely affected by future storms. In other words, by knowing the history of our coast, we'll have some idea of what's to come.

As early as 1925 those people mapping coastal changes had grasped this principle. Unfortunately, those who design coastal "protection" works did not reach the same understanding. In his 1925 letter to the captain of a Savannah steamer, R. S. Patton, chief of the Division of Charts, U.S. Coast and Geodetic Survey, wrote:

> It will be a long time before the basic principles are so well known that engineers can predict what is going to happen in any locality and design their structures accordingly. Until that time comes, about the best thing that can be done is to supplement what little imperfect knowledge of fundamentals we have by a study of what has happened in the past in the locality under consideration and then assume that the future will be in general a continuation of the past. The Coast Survey records comprise the largest collection of data extant with reference to these matters and I am sure that millions of dollars which have been literally thrown into the sea in coast defense work might have been spent as to serve some useful purpose if the Coast Survey records had been obtained and used intelligently.

Of course, coastal processes are not immutable through time. For example, if your grandfather owned an oceanfront house at East Beach on St. Simons Island, today you might be living in that same house, several blocks from the water. While he may have worried about erosion and vicious storms threatening to undermine his house, you might, on a lazy day, find yourself fretting about the long walk to the beach!

Knowing whether such changes reflect a true reversal of long-term trends, or are merely a temporary change in an ongoing cycle, requires a detailed knowledge of the area's history, a thorough understanding of complex coastal processes, and a crystal ball. The current state of the science is a rather broad delineation of hazard zones. Careful mapping of areas that were historically susceptible to particular natural hazards can indicate those areas potentially dangerous for development as well as those of lower risk.

This chapter aims to introduce the recent history of coastal processes on the Georgia shore. For example, do you know which inlets migrate and which are stable? Do you know which beaches are artificial and which are natural? Do you know which beaches are accretional (building up) and which are erosional (retreating)?

Recognition of the islands' dynamic nature holds important implications for the "best" and most appropriate use of these precious lands. Responsible development and management of island resources can avoid many natural hazards and conflicts of use. Wise assessment of past patterns, present trends, and future probabilities regarding shoreline changes and hazards should support such management and development decisions. We hope you will consider the economic and environmental benefits of appropriate land use and will encourage your political representatives to do the same.

Individual Island Analysis

The selection of a coastal property presents the prospective owner with two decisions: first, the wise selection of a safe and suitable *island*; second, the intelligent choice of a safe *site* on that island. The following guidelines for island selection, site selection, and site safety evaluation provide detailed directions for choosing an island property. As an additional aid to your research, the maps and accompanying descriptions at the end of this chapter present general analyses of each of the barrier islands along the Georgia coast. These broad island assessments are *not* designed to supplant your own site-specific evaluations! That decision—an essential one—is your own.

Selecting "Your" Island

You as a coastal dweller choose a style of life as well as a place to live when deciding to make your home on an island. As in selecting any location for a home, you must consider the aesthetics of the area; the convenience to work, schools, and stores; the cost of living; the political climate; and the availability of community services. You'll also need to consider some factors that affect the quality of life and the safety of both property and person.

Of the thirteen major barrier islands along the Georgia coast, only four currently have residential communities accessible by highways: Tybee, Sea, St. Simons, and Jekyll. Little Cumberland Island, which can be reached only by boat, has a small number of homesites, and availability of lots is strictly limited. The remaining islands, by various means, are protected from large-scale development and residential use.

Former barrier islands, such as Skidaway and Wilmington, are rapidly developing into heavily populated residential communities. Although these islands are often miles from the ocean, they are subject to changes and hazards that should concern island residents. For instance, the bluff at Isle of Hope was under 2 feet of water during a hurricane in the late 1890s!

While the socioeconomic characteristics of some of these islands may narrow your choices, the island you prefer should also be evaluated in terms of physical, biological, political, and infrastructure characteristics. All of the Georgia barriers are subject to the natural processes outlined in chapters 1 through 4. Life on any of them must accommodate the hazards of storm surge flooding, hurricane wind and waves, beach or river erosion, variable construction quality, and imprudent island land-use planning and development.

Each island is affected by such factors in greatly varying ways. In fact, there are variations in effects from site to site on a single island. Relatively "safe" sites can be found on islands generally unsuitable for development, while at the same time extremely high-risk sites exist on islands generally well-suited for human habitation. Choosing a home on the coast is *not* like shopping for an inland home. Choose your site carefully. Your safety and the survival of your property depend on it.

Physical Factors

On an Island
(1) *Proximity to waterfront.* How close is the site you're considering to open ocean, tidal inlet, riverbank, or marsh? The potential for property flooding and erosion is reduced at a site far from water, in the island interior, and at a suitably high elevation. Waterfront property, especially on beach or inlet shores, is likely to be eroding and susceptible to wind, wave, and storm surge hazards.

(2) *Elevation.* The extent of flooding by storm surges, extreme tides, or even heavy rains is primarily determined by elevation and the nature of nearby waters. For example, a lot with an elevation of 14 feet near the ocean may be more prone to flooding than a 10-foot-high lot near the sound.

The geologic age of the island, Holocene vs. Pleistocene, is often a determining factor in elevation—the Holocene (recent) islands being somewhat lower in elevation. The width and length of an island also should be considered in regard to flood hazards. The higher-elevation sites on a larger island, far from open water, offer relatively safe sites.

Check the elevation of the individual lot and the surrounding area with respect to the island's estimated storm surge and flood hazards. The highest elevations on the island are those safest from flooding, unless the property is adjacent to an area of erosion. U.S. Geological Survey (USGS) topographic maps show island elevations.

The flood hazard maps (Flood Insurance Rate Maps) issued by the Federal Emergency Management Agency (FEMA) predict the extent of expected flooding for storms of a given intensity. The FEMA flood hazard maps do *not*, however, take into account the probable effects of future erosion. If you live in an area of erosion, shoreline recession will eventually increase the hazard of flooding at your site. Let us repeat: current FEMA maps do *not* take into account these effects of erosion. (See chapter 7 and appendixes C and D for more information about the FEMA and USGS publications.)

Check the ages of both the topographic and the FEMA flood hazard maps. The 1957 topographic map series for the Georgia coast was updated in 1974. More recent orthophotoquads (elevation maps printed on a photograph base) will provide new and different information. The FEMA maps were revised in the late 1980s. Do not refer to outdated FEMA maps.

(3) *Soil type*. Soil type affects how well-supported and stable a structure will be. Also, the suitability of soil for septic tanks must be determined through testing. For example, a high water table or impermeable soil can result in septic tank failure. Check for soil profiles with layers of clay or peat and other organic materials (dark-stained or hard and impermeable layers). Such soils have a high water content that can cause problems for septic tanks; these soils also compact considerably and may not provide adequate support for houses or other structures. White-bleached sand overlying yellow sand to a depth of 2 to 3 feet often indicates stability because such a soil profile (often found in forest areas) requires a long period to develop.

The Coastal Georgia Regional Development Center (CGRDC) and the U.S. Department of Agriculture's Soil Conservation Service (listed in appendix C) have soil survey maps. Contact those agencies for information about your area. Further investigations and analyses of soil types should be conducted by experts.

(5) *Storms and hurricanes*. The probability of a hurricane striking should be a major consideration. Of critical importance is the ease of evacuation in the event of such storms.

Also, the history and severity of northeasterly winter storms in the island area should be taken into account. Check with long-time residents and local historical societies, as well as state and local coastal agencies, to find out what storms were like in the past. (Appendixes B and D provide information and references regarding hurricanes and storms on the Georgia coast.)

If your idyllic summertime vacation has inspired you to purchase an island home, it is imperative that you visit the site *during the stormy season*. Be there during the winter months: the beautiful wide beach that so enticed you in August may be a narrow strip of sand by December.

Visit during storms: you'll be amazed at the powerful effects of storm waves.

Be on hand during high tides and heavy rainfall; you may be surprised at how many buildings, high and dry during calm summer months, are regularly threatened by flooding. You may be surprised at how much sand is deposited in yards, streets, or driveways.

In short, unless you plan to own your property for only four months out of each year, you must know what your potential homesite looks like *year-round*.

(6) *Groundwater supply*. Groundwater is one of the most important natural resources of the sea islands. Fortunately, the Georgia coast is blessed with abundant high-quality water from the "Upper Floridan aquifer," stored in caverns, cracks, and crannies in underground limestones. In fact, some residents will remember free-flowing sea island wells (artesian). Recently, however, extensive (mostly industrial) use of the aquifer has reduced the "head," and marked cones of depression (areas of aquifer depletion) exist in some high-use areas such as Savannah, Brunswick, and St. Mary's (fig 5.1). Saltwater contamination of the aquifer has occurred at Brunswick and has become a concern in other areas of heavy use. For example, Hilton Head Island, South Carolina, has had its wells contaminated by saltwater as a result of pumping in the Savannah area.

Determine the status of the groundwater supply for the island in which you're interested. Has saltwater intrusion occurred near the island? Is the groundwater resource being properly managed? Are septic tanks threatening water quality? Agencies and resources listed in appendixes C and D can help answer these questions. In particular, the Water Resources Division of the Georgia Department of Natural Resources may prove helpful.

5.1 Savannah River industry, 1980. Heavy use of groundwater (mostly industrial) near Savannah and Brunswick has resulted in areas of aquifer depletion. Photo by L. Taylor.

On the Beachfront

All barrier island residents, especially those with beachfront property, should consider the following:

(7) *Sea-level rise*. Wherever sea level is rising, it may be caused in part by a rising of the world's oceans (called *eustatic* sea-level rise) and in part by local sinking of the land (called *subsidence*). This combined effect is called the *relative* sea-level rise for a particular area.

Subsidence rates for the Georgia coast are as great as 8 inches within the past century. Along the Georgia coast the rising sea level (0.01 feet per year from 1936 to 1975 near Savannah) also has aggravated the erosion problem.

If predictions of a "greenhouse"-induced acceleration in the rate of sea-level rise are correct, then island dwellers can expect greater storm damage and increasingly severe erosion in coming years. In 1990 a working group of the Intergovernmental Panel on Climate Change published a report hailed as the most broadly based assessment of the greenhouse threat conducted to that date (full citation in appendix D under *Sea-Level Changes*). The panel predicted that global sea level will rise between 3 and 11 inches by the year 2030.

(8) *History of shoreline changes*. Examine your property for signs of recent erosion (fig. 3.16). Evidence of an eroding shoreline includes (a) sand bluff or dune scarp along the upper beach; (b) stumps, mud, or peat exposed on the beach; (c) fallen trees or structures on the beach; (d) erosion control structures or renourished beach areas.

Study historical trends of shoreline changes in your particular area, with emphasis on the past 50 years. Talk to local geologists and planners. Note recent patterns of erosion or accretion. If your site is in an area of historic erosion, assume that trend will continue at a rate equal to or greater than in the past. If your site is in an area of historic accretion, determine, if possible, the reason for such growth and the probability of long-term accretion.

Areas near inlets are especially subject to extremely rapid cyclical changes—from erosion to accretion and back again. Consider most accreted areas near inlets to be ephemeral, especially on the north ends of islands and on the shores of smaller, extremely active inlets (for example, Cabretta). Historical patterns at most beachfront sites on the developed Georgia islands suggest continued erosion with continuing control efforts and increasing costs.

For further information on the history of shoreline changes, see the individual site analyses later in this chapter and the references in appendix D—in particular, the Griffin and Henry study cited under "Individual Islands." The state Coastal Resources Division also can give you information about erosion rates in your particular area (appendix C).

(9) *Size of the property.* The lot's depth should allow for suitable setback of structures from the edge of the ocean (or marsh or river). In particular, such setbacks are important when you consider the potential effects of sea-level changes and of continued beach (or riverbank or marsh) erosion. If you intend to move your house back as erosion continues, allow sufficient room for relocation.

(10) *Proximity of the site to an inlet.* Is the inlet dredged? Do channel jetties define the inlet? Patterns of erosion depend on factors implicit in these questions. Dredging of the Savannah River, for example, probably has increased erosion on Tybee Island by reducing the island's sand supply. Jetties disrupt the alongshore transport of sand. As mentioned, areas adjacent to inlets are subject to large, rapid, and sometimes unpredictable changes.

(11) *Dune system.* A well-developed system of primary and secondary dunes often signals a relatively stable or growing beach area. Keep in mind, however, that dunes are also subject to erosion (fig. 5.2a). You can see a well-developed natural dune system at Nanny Goat Beach on Sapelo Island (fig. 1.3), at the more accessible south end of Jekyll Island (fig. 5.2b), and on the spit of southern Sea Island. The dunes along the central portion of Tybee Island formed after the 1975–76 beach replenishment (fig. 5.2c).

How well-developed is the dune system in your area? Check the condition of the dunes. Are they building up or eroding? Also examine the extent and type of vegetation cover, the size of the dunes, and the width of the dune zone. More mature vegetation and higher, wider dunes are an asset.

Dunes are an important part of the sand sharing system. Remember that dunes are made up of windblown sand that originall came from the beach. By supplying sand to beach and offshore areas during storms, a dune system helps maintain the beach's na dynamic equilibrium (see chapter 3). The p ence of an extensive dune system also helps protect property and buildings from direct wave attack. The more dunes, the better.

5.2 (a) Dune scarp. A sharp, vertical "cliff" like this one near Goulds Inlet on St. Simons Island is a sign of recent erosion. 1989. Photo by T. Rust. (b) Well-developed natural dune field on Jekyll Island. 1989. Photo by T. Rust. (c) Dunes developed in the central portion of Tybee Island after the 1975–76 beach restoration project there. 1987. Photo by T. Clayton.

Selecting a Site on Georgia's Barrier Islands **61**

5.3 Small overwash fan, deposited by waves carrying sand into the marsh. North end of Jekyll Island, 1989. Photo by T. Rust.

(12) *Overwash*. Overwash occurs when storm waters carry sand through breaches in the dunes and into interdune and backdune areas (fig. 3.22). Sediments carried by the overwashing waters are deposited as fan-shaped masses of sand (fig. 5.3). On the Georgia coast, overwash primarily occurs on the ocean beach in low areas between and behind dunes. Along the riverbanks and on inlets and sounds, overwash of sand into marshes occurs.

Because the typical Georgia sea island is relatively wide, washover of the entire island rarely occurs except during major storms or hurricanes, and then mostly along narrow, low-lying areas such as spits. Complete overwash does occur at Little Tybee, Williamson, and Wolf islands.

On developed islands with no substantial dune system, overwash into yards and down beach access roads is not uncommon during storm tides. Structures in an overwash zone should be elevated to avoid blocking overwash flow. Sand removed from driveways and streets should be placed back on the beach.

On a Riverbank

Properties adjacent to marshes and tidal creeks or rivers, especially deep water, command high prices because many people want to own such scarce lots. Selecting this kind of property, however, usually on the backside or ends of an island, often involves a gamble with increased natural hazards.

These areas may not be as dynamic as some of the beachfront areas. Still, you should carefully evaluate the stability, safety, and suitability of the land for development. Especially consider the greater potential for erosion and flooding. If the property is right next to the river channel or a marsh with a tidal creek, you should take pains to investigate the factors listed above (including those in the *On the Beachfront* section). Pay special care to the following:

(1) *Erosion or erosion control structures.* Ideally, you should have an idea of shoreline changes that have taken place at the property that interests you. Historical data on "backside" erosion are generally unavailable, however, and the potential buyer is left to be his own detective.

Look for an eroding bluff or a bulkhead or riprap along the bluff of the property's upland. Signs of an eroding shoreline (fig. 5.4) include (a) a steep scarp along the bank; (b) fallen trees or structures along the bank; (c) erosion control structures or renourished beach areas. Take note of the position of any nearby rivers or creeks. Watch especially for meanders, or bends, in the river's course. The outsides of these bends are areas of active erosion as the river changes course. If your property is on the outer edge of the curve of a meander, you can expect to lose some property, sometimes quite rapidly, as a result of continued erosion.

Often the cost of preventing such erosion is high. Construction of timber bulkheads can cost more than $200 per foot. Also inspect adjacent properties to determine which, if any, methods of erosion control are being used. Again, consider the size of the lot. Is the property deep enough for an adequate setback from the water's edge? Is there enough room to relocate threatened structures?

(2) *Elevation.* Study the FEMA flood hazard maps and USGS topographical maps to determine the property's elevation and potential for flooding. Along the marsh edge of the property, inspect the vegetation for any upper marsh species that may delineate an upper zone of flooding during spring tides. In addition, inspect low-lying areas for wetland vegetation that may indicate frequent flooding from rainfall. (The field guides suggested in appendix D can help you identify upper marsh vegetation.) Remember that FEMA regulations

require builders in communities covered by flood insurance to adhere to construction guidelines for high hazard or flood zones, even if the property is on the island's marsh or river edge.

(3) *Presence of dredge spoil fill.* If the soil on the property contains coarse (large-grained) sediments, almost gravel-like sands, with a

5.4 An eroding riverbank. Signs of erosion include the steep scarp and the fallen trees. Cumberland River, 1985. Photo by L. Taylor.

5.5 Finger canal on Tybee Island, 1989. Photo by T. Rust.

mixture of some clay masses and black-and-white shell material, then the area may consist of dredged materials. (However, don't confuse oyster shell-laden Indian midden sites with dredge spoil areas.) Pumped from river bottoms or brought in by truck, this fill material may have been placed on top of marsh land. Such soil may not provide proper support or stability for construction. If you decide to locate on such material, make sure there are no buried peat beds.

(5) *Finger canals.* A channel excavated on the backside of an island (fig. 5.5) for the purpose of creating waterfront lots is a source of potential problems. The worst of these are (1) lowering of the groundwater table, (2) pollution of groundwater caused by salty canal water seeping into the groundwater table, (3) poor tidal flushing and stagnation, (4) fish kills resulting from high water temperatures, nutrient overloading, and deoxygenation, and (5) canal water polluted by septic effluent. Such canals are relatively rare in Georgia.

Biological Factors

(1) *Dune vegetation.* The degree to which dune vegetation has proceeded through stages to more mature, more stable species indicates the dune system's relative age. Older and more stable dunes (fig. 5.6) usually indicate that a longtime accretionary or stable beach exists.

Sparse vegetation on dunes may indicate the presence of livestock (cattle, horses) or human beings who destroy plant life. Dune growth is inhibited when stabilizing vegetation is destroyed, and dunes may be reactivated, resulting in the nuisance of blowing sand that may damage surfaces or migrate onto driveways, steps, and streets. Destabilization of dunes also has had negative effects on some of the "undeveloped" barriers.

(2) *"Green zone."* A "green zone" of trees and shrubs acts as a setback area for structures built near the beach. Structures with adequate setbacks are better-protected from storm energies and the potential effects of erosion.

Such a setback zone may be evidence that the community is aware of the importance of distancing structures from the dynamic area of the beach. Or it may simply be that no one has gotten around to developing that area yet, and the green zone is merely accidental. Officers'

5.6 Mature dune vegetation and stabilized dune system. Note the award-winning dune walkover constructed to restrict foot traffic and protect the dunes from devegetation and destabilization. Holiday Inn, Jekyll Island, 1987. Photo by T. Clayton.

Row on Tybee Island, for example, is adequately set back from the oceanfront. However, in 1991 houses were being constructed in the former green zone, *seaward* of the row.

If setback areas do exist, learn what laws establish the setback and whether state and local managers have the authority, commitment, and resources to maintain the integrity of the green zone. The state's Shoreline Assistance Act establishes setback lines for construction; however, the act does not promote the creation of a green zone of any standard width or depth.

The setback zones also should incorporate projections for the effects of sea-level rise during the life of the structure. Although some other states have devised laws or regulations to take account of anticipated future changes, Georgia has not.

(3) *Maritime forest vegetation.* If your house is strategically nestled within the maritime forest, the forest vegetation provides further protection from the effects of storm waters, winds, and erosion. However, a maritime forest alone is not sufficient to indicate long-term stability; any land next to the beach or marsh may be subject to severe erosion (figs. 3.16b and 5.4).

Also make sure that the forest around you will remain forest. Just because you recognize its value doesn't mean that your neighbors will. Few extensive stands of maritime forest remain on Georgia's developed barrier islands, but you can still see marvelous examples on Wassaw and Blackbeard Islands; relatively undisturbed forest stands also exist at the more accessible northern ends of Sea and Jekyll Islands.

(4) *Salt marshes.* For homes on the island's backside, notice the extent of the marshes adjacent to the island. Look for any tidal creeks that may be potential sources of bank erosion. If the outside curve of a tidal creek is adjacent to an island bluff, the creek waters may eventually erode the upland as the creek meanders.

Political and Infrastructure Factors

(1) *Federal, state, county, and local regulations.* Island dwellers should pay particular attention to those regulations regarding flood hazard mitigation, flood insurance, growth management, zoning and land use, building construction, site location, setbacks, and erosion control. Chapters 6 and 7 provide some of this information; discussions with local government officials should fill in the rest. The potential homeowner also should

find out whether regulations on the books are being enforced.

(2) *Zoning and land-use ordinances.* To avoid purchasing property in an area which may eventually be developed in ways incompatible with your desires, determine the property's zoning classification. Keep in mind, however, that zoning laws are not immutable but subject to variances and modification. Also, check the island land-use plan to determine the community's anticipated growth (patterns and magnitude). Study how buildings in the community are constructed and talk with area planners to learn whether ordinances, regulations, and codes are properly enforced. Taking an active part in community affairs is in your best interest, particularly when these issues are involved.

(3) *Patterns of island use.* Determine the differences in the year-round and seasonal populations. Is the island too crowded during the summer season? What is the maximum population density for the island? Are plans in place to ensure that this maximum population is not exceeded? What is the size and nature of the voting population? Will you have a voice in determining the direction of future development? What is the philosophy of public officials and the business community toward island development? Discussing such issues with area planners and citizens' groups can help answer these questions.

(4) *Community services.* Does the island community have adequate water supplies, solid waste and sewage facilities, fire and police protection, schools, and highways? (For example, is the local fire department capable of handling high-rise fires? Is there sufficient water pressure to combat such fires?)

Evaluate the island's land-use plan. Determine whether the county or local government has prepared a long-range plan for managing the island's resources and growth. If a suitable plan is in place, does the municipality adhere to its guidelines?

Pay attention to nearby industries that may contribute to air, water, soil, and noise pollution or contamination. Growth management on an island is extremely important to avoid overwhelming the limited resources of the island and the local government.

(5) *Evacuation.* Determine the suitability and location of evacuation routes from the island. How far is the nearest storm evacuation route from your property? How will the large summer population affect your ability to evacuate your family quickly in the event of a storm or hurricane? Note the elevation of the causeway and bridge that connect the island to the mainland. Low causeways or bridges will be quickly flooded, and escape routes from the area will be closed. On Tybee Island, for example, the island roads are on high ground and most bridges are high. However, a few areas of the mainland-to-island causeway (fig. 5.7), across which everyone must escape, are low and often flooded during exceptionally high tides.

5.7 Causeway to Tybee Island, 1989. Long, low causeways are potential evacuation hazards because of their susceptibility to early flooding. Photo by T. Rust.

Also, realize that drawbridges may malfunction during an evacuation. In fact, it is best to assume that drawbridges *will* fail to operate. High-span bridges, while safe from flooding, are unsafe in high wind conditions and will be closed once wind speeds reach a certain threshold (which varies from bridge to bridge).

The CGRDC has developed evacuation plans and maps for the populated coastal areas; they are published in your local phone book. Contact the CGRDC or local Civil Defense personnel with questions about evacuation.

(6) *Erosion control structures.* Erosion control structures indicate severe erosion problems. If a site (or an adjacent site) is protected by a seawall or other stabilization device, severe erosion problems already exist. You should expect these problems and costs to increase with time and continued erosion.

Study the history of erosion control efforts for your island, especially for the beach area fronting your property. Even if no erosion control structures are visible, they may have been extensively used in past years. Beach replenishment often covers evidence of former control structures. However, such structures may become exposed as the artificial beach erodes—as on Tybee Island. Determine how erosion control structures and beach replenishment are paid for and what your personal contribution will be.

Tables 5.1, 5.2, 5.3, and 5.4 provide the histories of erosion control efforts on the Georgia coast, and appendix D will guide you to further information.

Site Safety Checklist

The following checklist recaps characteristics essential to site safety.

—The site and neighboring areas are above projected storm-surge levels.
—The site is behind a protective natural barrier (that is, dune system) and away from overwash points such as access roads cut straight through dunes to beaches.
—The property is not adjacent to a migrating inlet. The property is not within reach of cyclical inlet changes.
—The site is not in an area of shoreline erosion along a beach, river, or creek.
—The site is located on a portion of the

Table 5.1 History of Erosion Control Efforts at Tybee Island

Date	Structure	
1882	3 groins	Probably by the Army along north shore of Fort Screven. Repaired occasionally until 1886. Destroyed by 1939.
1903	Repairs to Fort Screven seawall	Unclear whether this planned $4,000 repair was funded and carried out.
1904	Repairs to Fort Screven seawall	Unclear whether this planned $5,000 repair was funded and carried out.
1904–1906	Riprap	By Army at Fort Screven.
1907	Seawall (4,000 ft)	By Army at Fort Screven. $50,000. Repaired 1908.
1908	Repairs to 1907 seawall	By Army at Fort Screven. $19,000.
1908	Groin; seawall additions	By Army. $1,000.
1909	9 groins (or more)	By Army at Fort Screven. $5,200.
1912	Spur dikes/4 groins	By Corps at Fort Screven.
1917	5–6 groins	By private interests at south end of island. Destroyed by 1939.
1919	5–14 spur "jetties"	Unclear whether these Corps groins were built.
1921	10 groins	By Town of Tybee between Tilton Avenue and Fourth Street. "Completely disappeared" by 1930.
1922	5 groins	By Corps at Fort Screven.
1922	Groins	By Town of Tybee.
1929	Seawall	By Town of Tybee between Tilton Avenue and Third Street.
1929–1930	4 groins	By Town of Tybee between Tilton Avenue and Third Street.
1930	2 groins	Two temporary experimental groins built by Army at Fort Screven in April 1930. Repaired twice. Destroyed by May 1, 1930.
1930–1931	Seawall	By Corps at Fort Screven.
1930–1931	Riprap	By Corps at Fort Screven.
1930–1931	5 groins	By Corps at Fort Screven. Built seaward of 1930 seawall and landward of remains of 1907 seawall. $85,000.
1932	Groin extension	By Corps at Fort Screven. To repair groin breached by storms of March 1932.
1932	Riprap	By Corps at Fort Screven. Total cost of groin extension plus riprap, $1,600.
1933	Repairs to Fort Screven 1931 seawall	By Corps. To repair 122-foot section of 1931 seawall that failed during November 1932 storm.
1933	Seawall	Fort Screven 1931 seawall extended 1,070 feet north by Corps.

Table 5.1 (continued)

Date	Structure	
1933	Groin extension	By Corps at Fort Screven.
1933	4 groins	By Corps at Fort Screven. Total cost of 1933 construction, $66,000.
1936	Seawall and 19 groins	By Works Progress Administration (WPA) from south end of island to Tilton Avenue. $93,500. By 1939, groins along southern half had been destroyed and rebuilt 3 times.
1936	Riprap	By Army at Fort Screven. $700.
1937–1938	Seawall	By WPA at Fort Screven.
1937–1938	5 groins	By WPA at Fort Screven.
1938–1939	Seawall	By WPA between Fourteenth Street and south end of island.
1938–1939	9 groins	By WPA between Fourteenth Street and south end of island. 6 replacement groins, 3 new groins. $209,000.
1939	4 groins	By WPA along Savannah Beach.
1940–1941	Repairs	By Corps at Fort Screven. $6,000.
1941	Seawall	By WPA between First Street and Tybee Creek.
1941	19 groins	By WPA between First Street and Tybee Creek.
1946	Dike/seawall	By Town of Tybee at Fort Screven. Destroyed in gales of 1962 and February 1963.
1965	Riprap	By Federal Office of Emergency Planning from Tybee Light to Seventh Street. Placed to reinforce seawall in wake of 1964 Hurricane Dora erosion landward of wall. $301,000.
1971	Seawall/riprap	By Federal Office of Emergency Planning. Emergency repairs to 100 feet of seawall that collapsed during storm of November 1971. $65,000.
1974–1975	Terminal holding groin	By Corps near Fort Screven to contain artificial beach.
1975–1976	Beach replenishment	By Corps from 1974–1975 groin to Eighteenth Street. $2,600,000.
1984	Riprap	Emergency protection for seawall south of Seventeenth Street. $321,900.
1987	Terminal holding groin and groin repair	By Corps. New groin at south end of island built to contain Corps' artificial beach. 1975 groin at north end of island repaired. $606,220.
1987	Beach replenishment	By Corps at south and north ends. $1,989,000 for 1 million cubic yards.

Source: Modified from *A Study of Beach Erosion and Beach Erosion Control Efforts on Tybee, Sea, St. Simons, and Jekyll Islands, Georgia, and Amelia Island, Florida*, by L. Taylor.

Table 5.2 History of Erosion Control Efforts at St. Simons Island

Date	Structure	
pre-1940	Seawall	By private interests along south shore of island to a point west of the county pier.
pre-1960	Revetment	By private interests.
1964–1965	Revetment	By Federal Office of Emergency Planning in two sections: (1) from King-and-Prince Hotel to 1,900 feet west of the municipal pier, (2) from Tenth Street to 3,000 feet north.

Source: Modified from Taylor, *A Study of Beach Erosion*.

Table 5.3 History of Erosion Control Efforts at Sea Island

Date	Structure	
?	Revetment	By Sea Island Company in vicinity of Beach Club.
1964	Revetment	By Sea Island Company in vicinity of Beach Club to replace earlier revetment destroyed by Hurricane Dora.
1965	Beach replenishment	By Federal Office of Emergency Planning. $170,000.
1964–1968	Minor beach replenishment	By Sea Island Company in vicinity of Beach Club.
1972	Revetment	By individual homeowners.
1975	Revetment	By individual homeowners.
1977–1981	Revetment	By individual homeowners along entire beach backed by private homes. Built in response to severe erosion that began in 1976–1977 on north end near Thirty-Sixth Street.
1981	Revetment	By Sea Island Company from Fourth Street to south of First Street.
1981	Gobi Mat	By Sea Island Company from south of Harrington Street to south of First Street.
1981	Revetment	By Sea Island Company for 0.7 miles north of Thirty-Sixth Street.
1987	3 groins and minor beach replenishment	By Sea Island Company in vicinity of Beach Club.
1990	Beach replenishment	By Sea Island Company.

Source: Modified from Taylor, *A Study of Beach Erosion*.

Table 5.4 History of Erosion Control Efforts at Jekyll Island

Date	Structure	
pre-1964	Seawall	In vicinity of north bathhouse. Partially destroyed by Hurricane Dora, 1964.
1964–1965	Riprap/seawall	By Federal Office of Emergency Planning. 2 sections: (1) south of Aquarama, (2) in vicinity of north bathhouse. $192,000.
1964–1974	Riprap/seawall	By Jekyll Island State Park Authority. 2 sections: (1) in vicinity of south bathhouse, (2) at north end of island.

Source: Modified from Taylor, *A Study of Beach Erosion*.

island backed by extensive salt marsh.
—The site is located away from low, narrow portions of the island.
—The site is located behind or within an area of mature maritime forest that will not be removed for future development.
—The site readily drains water.
—Fresh groundwater supply for the area is not threatened by saltwater intrusion, contamination, or depletion.
—Soil and elevation are suitable for septic tanks (if necessary). Septic systems and water wells are properly spaced. Sewage treatment facility functions properly and within capacity.
—The soil profile contains no compactable layers such as old marsh peat in the area of the footings.
—Structures on adjacent properties are adequately spaced and soundly constructed.
—Access to storm evacuation routes is quick, easy, and free from early flooding.

The remainder of this chapter provides information specific to portions of each island. Additional materials in the form of maps, aerial photographs, scientific literature, and planning documents contain much valuable information too extensive to include in these pages. The bibliography in appendix D lists such materials. Also, discussions with local planners, administrators, scientists, and activists can provide valuable insight into the structure, operations, and character of each island and community.

Table 5.5 Summary of Conditions Considered in the Assignment of Risk Classifications

Extra high risk (XH)
Elevation averages <5 feet
Flooding occurs frequently in association with spring tides
Dune system is nonexistent
Continuous, relatively rapid erosion of shoreline, especially since 1924 shoreline survey; continuous since 1855
Long history of engineered attempts to control beach erosion; dysfunctional groins present; seawall/bulkhead present at MHW shoreline; no renourishment planned
Adjacent to an actively migrating inlet or tidal creek
Frequent dune overwash during storms

High risk (H)
Elevation is generally low, 5–10 feet
Flooding occurs infrequently on spring tides; more frequently on storm tides
Dune system is poorly developed or eroding seriously
Trend of net erosion since 1855
Current use of erosion control structures with active plans for renourishment of the shoreline
Adjacent to a currently inactive, relatively stable inlet with a history of a tendency to migrate
Infrequent overwash associated with major storms

Moderate risk (M)
Average elevation is 10–15 feet
Flooding occurs only during storm tides
Dune system is an immature system, perhaps with some erosion
Stable, or slightly accretional, patterns of shoreline change since 1924
Infrequent use of engineering structures along the shoreline; no current need for renourishment
No recent trends of inlet migration
Overwash is unlikely except during breaching by hurricane storm surge waters

Low risk (L)
Average elevation is >15 ft.; site is well inland
Flooding is likely only with a storm surge of a major hurricane
Dune system is well-developed/mature; not eroding
Continuous accretion along shoreline since 1924
No history of or need for engineering structures or renourishment
Not adjacent to an inlet or tidal creek

The Georgia Islands

The Rating System

Developers, realtors, and sellers of island property—commercial and residential—ideally should feel responsible for providing information about possible hazards and probable levels of risk. Be aware, however, that the buyer, not the seller, must make such determinations.

The following maps and text offer general comments on hazard status over large areas. Characteristics of a particular individual site may therefore differ dramatically from the rating zone's overall characteristics. Only by thorough evaluation of an individual site can specific site suitability be determined.

The reader also will note that some islands are discussed in greater detail than others. For those islands presently owned and managed by conservation-oriented families, foundations, or agencies, only a brief introduction to the island is presented. These islands are unlikely to soon face most of the issues discussed in this book. Thus, already developed or developing islands will receive most of our attention.

The primary factors considered in rating lands adjacent to a beach, marsh, or creek shoreline include (1) average elevation of the upland and condition of the dune system; (2) "historical" (1855–1924) and "recent" (1924–1974) net changes in mean high water (MHW) shoreline position; (3) history of erosion control efforts (groins, seawalls, renourishments) along the shoreline; (4) historical and potential effects of inlet migration; (5) potential for storm overwash.

Based on the type and severity of potential hazards (table 5.5), "extremely high risk," "high risk," "moderate risk," and "low risk" areas are designated for island segments. The interpretation of hazard potential for each island segment sums up the information in scientific and planning documents, analyses of maps and aerial photographs, and field observations and discussions.

Of course, despite relying on the best available data for the rating system and analyses of each island, interpretation of this information is necessarily subjective. The rating system is weighted to favor long-term trends and histories of shoreline change in making evaluations. Seasonal conditions and episodic events, including beach nourishment, may change the character of a specific site in the short term.

Tybee Island

Tybee, the northernmost Georgia barrier island, is 750 to 1,500 years old (figs. 5.8 through 5.17; table 5.6). Its name comes from a Native American word for "salt," which is appropriate for this sandy, scrubby island.

The island is 3.5 miles long and 2.5 miles wide (including adjacent marsh); the average elevation is less than 15 feet. Located along the southern margin of the Savannah River, Tybee is part of a geological unit that includes the Little Tybee Island complex, Williamson

Table 5.6 Tybee Island Shoreline Hazard-Risk Evaluation

Shore-line section	Comments
1	*Lazaretto Creek to northwest end of river beach* High salt marsh and former landfill site Flooding likely during high tides and storms Historical accretion; recent stability No engineering structures Minor migration of inlet
2	*Northwest end of river beach to Lighthouse Point/Van Horne Drive* Flooding/overwash unlikely except during major storms Recent (since 1925) shoreline position constant with historical (1866–1925) accretion Few erosion control structures
3a	*Lighthouse Point to north end terminal groin* Flooding/overwash of low dunes during storms Long-term trend (1866 to present) of cyclical accretion and erosion; MHW shoreline relatively stable; some accretion since 1970s
3b	*North end terminal groin to Center Terrace* Flooding/overwash behind dunes occurs during storms; poorly developed dunes Slight accretion since 1925; erosional since 1866 Long history of erosion control efforts
4	*Center Terrace to Fourteenth Street North* Well-developed foredunes and backdunes; overwash only during major storms Stable to accretionary shoreline since 1925; accretion since 1866 Erosion control efforts include groins, seawalls, and beach replenishment
5	*Fourteenth Street North to Tybee Creek shoreline* Infrequent flooding of shorefront property during storms and extremely high tides Recent trend of erosion (since 1925) Extensive history of erosion control efforts Migration of channels through ebb tidal shoals affects patterns of erosion and accretion Dunes absent except for dune restoration project efforts
6	*Tybee Creek shoreline* Minor flooding associated with dune overwash during storms Slight net accretion since 1925; net erosion since 1866 Dysfunctional groins and bulkheads along shoreline Discontinuous dunes
7	*Venetian Drive marshfront shoreline* Infrequent flooding of road and adjacent yards Migrating inlet at Horsepen Creek, with recent trends of accretion on Tybee Some bulkheads along edge of apparent fill

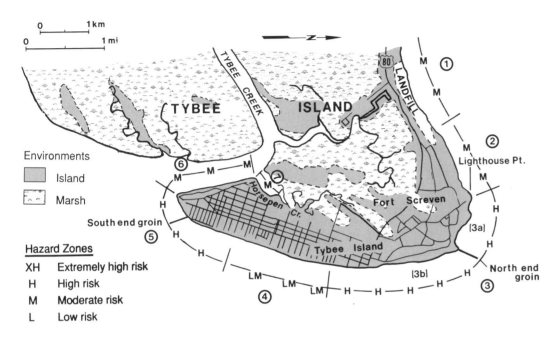

5.8 Site analysis map: Tybee Island. Companion map to table 5.6.

Island, and adjacent marshes (fig. 3.12).

Use of the island as a site for navigation aids and military posts continued from colonial times through the mid-20th century. The first limited use of Tybee as a recreation area occurred during the early 1800s, but it was not until the opening of the railroad in 1887 that the island began to grow into a truly popular beach resort.

The construction of a highway from Savannah to Tybee in 1933 ushered in the next era of accelerated beach use and land development, especially at the south end of the island. Even today this area of Tybee (fig. 1.1b) remains the focal point for many island visitors because of its easy beach access and concentration of businesses and recreational facilities.

Use of Tybee's northern end as a military reservation involved the U.S. Army in an unceasing and losing battle with the sea—an opponent more formidable than any military force that might seek to invade Savannah. Tremendous losses of land occurred between 1868 and 1924. Gun emplacements, observation towers, parade grounds, military housing, and erosion control devices were lost to encroaching ocean waters.

Since the sale of the federal property to the public after the military reservation's closing, private homes have been constructed on the north end beaches. Today, several houses are located on the back of the dune line.

Although Tybee's growing population during the mid-1900s did prompt housing development along the oceanfront, many structures were set back from the beach along the island's middle and northern portions. Unfortunately, structures built in this area since the 1960s have been placed much closer to the beach. Fortunately however, shoreline changes along Tybee's central portion have not been severe, and only a few streets and structures have been lost.

Although actual shoreline changes have been greatest on the island's north end (fig. 5.9), beach erosion problems seem to be more severe at the south end because of the relatively heavy concentration of development crowding the shore there.

Elevation. The highest areas on the island are approximately 16 feet above mean sea level (MSL). Some of the island's highest points include the ridge crests behind the main island and parallel to the Savannah River inlet shoreline, the crests of some active backdunes, and the area along Butler Avenue near Twelfth to Fourteenth Streets. The remainder of the

Selecting a Site on Georgia's Barrier Islands 71

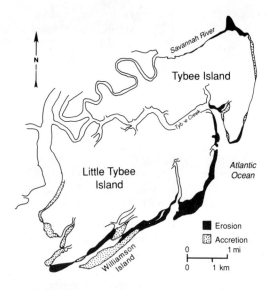

5.9 Historic net erosion and accretion: Tybee Island and Little Tybee Island complex, 1925–1974. (Adapted from *Historical Changes in the MHW Shoreline of Georgia, 1857–1982*, by M. Griffin and V. Henry, 1984. Full reference given in appendix D under *Individual Islands*).

inland area is lower than 12 feet above MSL. Some areas along the edges of the tidal marshes are subject to flooding during high spring tides. Backdune heights range from 10 to 14 feet along the central beaches (from Sixth to Thirteenth Streets), down to 4 feet within the newly formed dunes of the restoration project. During the "great hurricane" of 1893, eyewitnesses noted that some of the island's highest points were under 6 to 8 feet of water.

History of MHW Shoreline Changes. Because of Tybee Island's proximity to the Savannah River, shoreline changes on the island are linked to navigation and flood control activities on the river—such as the dredging of channels, construction of upstream dams, and institution of upstream soil conservation measures—all of which remove sediment from the river system.

The extent to which these activities affect the Tybee shore is controversial. The Corps of Engineers, which conducts many of the river alterations, has officially maintained that "there is no evidence that the Savannah River jetties or channel maintenance have affected the normal drift pattern."

However, the Georgia Geologic Survey study of historical maps and photos by Griffin and Henry reveals two distinct trends in Tybee Island's shoreline history. Prior to 1913 net accretion was evident. After 1913, however, when river dredging began on a regular basis, the shoreline experienced net erosion. Although attributing this change to navigation and flood control projects is not unanimous, it should be no surprise that removal of millions of cubic yards of sediment from the nearshore river/island system would have a noticeable impact on Tybee's shoreline.

From 1866 to 1974 the *net* change in MHW shoreline position was relatively small for many areas; that is, the 1866 shoreline position did not differ much from that of 1974. The greatest erosion occurred on the northeast oceanfront, while some erosion took place along the Tybee Creek inlet shoreline. Accretion occurred along almost all of the Savannah River inlet shoreline and the central and southern ocean beaches.

Tybee did not lengthen or move to the south during the study's 108-year period. As Griffin and Henry write, Tybee "maintained a tenuous dynamic stability." Dynamic, indeed! Dramatic changes in shoreline position occurred, especially on the northern Atlantic Ocean and Savannah River shores. Although the net difference between the 1866 and the 1974 shorelines may not be that great, migrations of the MHW line were impressive.

In the 50 years from 1925 to 1974, net shoreline erosion was greatest on the Savannah River inlet shoreline, on the southern ocean beaches, and along the Tybee Creek shoreline (fig. 5.9). The remainder of the beaches underwent little or no accretion.

North Tybee Island

The northern shoreline of Tybee Island (facing the Savannah River) has undergone net accretion over the past 120 years; however, the shoreline here also has experienced periods of recession, most notably from 1925 to 1957. Between those years, the north end spit in fig. 5.9 disappeared at rates approaching 60 feet per year.

Much of the sand responsible for the most

recent episode of accretion (not shown on these maps) is "runaway" sand from the Corps' 1975 oceanfront artificial beach. This sand passed over and through the leaky north terminal groin and was deposited on the Savannah River shore (fig. 5.10). This recently accreted sand, therefore, was a one-time, fortuitous infusion and not part of any long-term natural change or pattern. If the 1987 repair of the north groin proves successful, the north shore can no longer look to the beach nourishment project as a source of sand.

The older houses of Officers' Row are sited on a high, continuous ridge. Considerable accretion has occurred seaward of the homes, and this new land provides a significant buffer from erosion and wave attack. Any structures seaward of Officers' Row will be built at a significantly greater risk.

Shoreline erosion in the area of the Butler Street/Highway 80 intersection has threatened the roadway and several structures north and south of the turn in the road (figs. 5.11a, 5.11b). A seawall with rock rubble armoring the base has been emplaced to reduce the upland erosion rate. Although protected immediately after beach replenishment, this area is threatened by ocean waves and flooding when erosion exposes the seawall again (fig. 4.12).

Central Tybee Island

The central portion of the island (fig. 5.12) has been relatively stable (1925–1975). "Stability" in this case, however, does not mean there has

5.10 North end of Tybee Island. The large terminal groin (arrow) marks the northern boundary of the beach replenishment project. The bulge of sand on the Savannah River shore is the "runaway" sand from the 1975 beach restoration. Early 1980s. Photo by W. Cleary.

5.11 Tybee Island, in the vicinity of the Highway 80/Butler Avenue intersection. (a) Top left: Note that there exists no high-tide beach here, despite the Corps of Engineers' replenishment project. Early 1980s. Photo by W. Cleary. (b) Top right: Flooding behind the seawall, October 1985. This home was only a few years old when this photo was taken. Photo by T. Clayton.

5.12 Central Tybee Island. Note the protective dunes between the shoreline and the first row of houses. Early 1980s. Photo by W. Cleary.

been no change in shoreline position. Rather, *cycles* of erosion and accretion have occurred, adding up to little or no *net* change.

For example, the strand area of Tybee Island experienced accretion of 2 to 20 feet per year between 1866 and 1913, erosion of up to 50 feet per year between 1913 and 1925, continued erosion averaging 6 feet per year between 1925 and 1957, accretion of 200 to 400 feet between 1957 and 1974, and rapid erosion of up to 22 feet per year between 1974 and 1982.

Therefore, beware of statements about *average* or *net* erosion rates or changes. If your house happens to be sited on land routinely removed or threatened (albeit temporarily) by erosive cycles, it is little comfort to know that in another few years your lot will be returned by an accretionary cycle.

South Tybee Island

The southern end of the island (fig. 5.13) has experienced large fluctuations in shoreline position. In 1867 the shoreline was 1,200 feet *landward* of the present seawall. Yet in 1900 the shoreline was 3,500 feet *seaward* of the same seawall.

This area is primarily affected by the behavior of the Tybee Creek tidal channels. Another variable introduced into shoreline behavior in this area is the dredge and beachfill project being coordinated by the Corps of Engineers.

The most recent major change was the rapid 1975–76 disappearance of an initially wide beach, caused perhaps at least in part by

5.13 Southern tip of Tybee Island, with seawall and groins. November 1980. Photo by W. Cleary.

5.14 (a) Storm waves breaking over seawall and onto Tybee Island south parking lot, 1981. Photo courtesy of U.S. Army Corps of Engineers, Savannah District. (b) The same parking lot, after the 1987 beach replenishment. February 1988. Photo by T. Clayton.

changes in Tybee Creek Inlet. These changes may have been triggered by the Corps' dredging during the beach replenishment project or they may have been caused by natural changes in the inlet shoals (see chapter 4).

The most severe erosion was in the section south of Fourteenth Street in the island's most heavily used beach area. By 1981 there was no high tide beach between Sixteenth Street and a point south of Eighteenth Street. During storms or spring tides, waves were actually breaking over the top of the seawall, onto the parking lot, and onto the porches of three houses south of Eighteenth Street (fig. 5.14a).

In 1987 these houses and the seawall were granted a temporary reprieve when the Corps pumped sand ashore to re-create the formerly wide beach (fig. 5.14b). But by 1991 the south end groin remnants and seawall were again exposed; several swimmers were injured during the summer tourist season. Although protected immediately after a new beach is pumped up, this area is threatened by ocean waves and flooding when erosion exposes the seawall again.

The southwestern shore of Tybee Beach faces Tybee Creek. This area is relatively stable and even accreted at rates of 12 to 25 feet per year between 1974 and 1982. The proximity to Tybee Creek does introduce an element of uncertainty for the future of this section.

History of Erosion Control Efforts
Because Tybee Island affords an opportunity to see how the New Jerseyization of a Georgia island can occur, it is worthwhile to take a detailed look at the history of stabilization there.

The whole of oceanfront Tybee Island has had a long history of fighting the sea (table 5.1). The battle began more than a century ago; in 1855 a mapping project was begun in response to erosion problems, and by 1882 groins were already being placed at the island's north end to protect threatened gun emplacements. In the early 1900s when Fort Screven was an active military installation, the federal government found itself endlessly battling the ocean.

Historical records are skimpy, but one of the earliest major erosion "control" efforts was probably the 4,000-foot "Old Seawall" built by the Army in 1907 from the north end of the island to First Street. In 1928–29 a series of groins (the Central Beach Protection Works) was built between Tilton and Third Streets.

The 1930s brought the Work Projects Administration (WPA) onto the scene. By 1931 the "Old Seawall" had apparently failed quite dramatically, for the new WPA seawall and groins were built *behind* it (i.e., landward). Within two years, part of this new half-mile seawall had failed, and plans were made to repair and reinforce it. However, funding problems and shifting erosion patterns delayed that effort and shifted attention to the northernmost end of Fort Screven. By 1938 an additional 0.4 mile of seawall and five more groins had been added to the Savannah River shore.

The ocean, not surprisingly, continued to defy those who insisted on their own definition of the shoreline's position. The next few years, 1939 to 1941, sent the seawall builders scurrying to the island's central and southern portions.

The initial WPA project of 1939–41 called for a seawall and seven groins in the area from Tilton Street to Second Street. This plan was altered to extend the seawall down to Fourteenth Street and to add five more groins plus a small beach replenishment project at Tybee Creek Inlet. Two concurrent WPA seawall projects were built to "close the gaps in the general protection scheme for the entire island," thereby encasing almost all of oceanfront Tybee in a shell of armor.

These projects continued to tinker with the existing seawalls, changing angles here, building more groins there: symptoms of advancing New Jerseyization. Many groins were built in conjunction with the seawalls. In all, more than 120 groins have been built along the Tybee Island oceanfront in the past century. Although most of these structures have been destroyed, a number of them are still exposed (fig. 5.13), cluttering the beach and endangering swimmers (figs. 4.1, 5.15a, 5.15b).

In 1946 when Fort Screven became part of the Town of Savannah Beach (now the City of Tybee Island), the municipality continued the battle. In 1962–63, however, when storms destroyed the dike that had been built and eroded several hundred feet of land behind the seawall, town officials realized they were overmatched. Subsequent appeals to the Corps of Engineers brought the federal government back into the ocean-battling game. In 1964 Hurricane Dora made landfall at Tybee Island and severely damaged the seawall between First and Center Streets. The Federal Office of Emergency Planning paid for the Corps to place almost a mile of stone riprap to provide emergency reinforcement for the damaged wall.

Let's step back for a moment from this dizzying flurry of "erosion control" construction and look at the overall picture. As of 1970, after more than a century's experience of studying, charting, surveying, constructing, patching, extending, reconstructing, and reconstructing again "erosion control" works, what was the end result?

A Georgia Sea Grant study published in 1972 found that (1) although the seawall prevented the loss of the private property behind it, the erosion continued to lower the shoreface

5.15 South end of Tybee Island, same area as in fig. 5.14. What have we done to our beaches? Early 1980s. Photos by W. Cleary.

seaward of the wall (thereby destroying the public's recreational beach); (2) the groins were ineffective at trapping sand and inhibiting erosion; and (3) some groins apparently formed rip currents that actually may have increased local erosion.

The end result, then, is a shoreline cluttered with ineffective or even damaging erosion control works. In addition, the structure that was performing most "satisfactorily" was the seawall protecting the upland property, but at the expense of the public's recreational beach. And all this mostly at taxpayer expense!

In 1970 the Corps recommended to Congress that the federal government help pay for a new artificial beach at Tybee Island to replace the natural beach destroyed by the seawall. In 1975–76 federal, state, and local taxpayers paid approximately $3 million to pump up 2.6 million cubic yards of sand along the oceanfront. But the new artificial beach did not behave exactly as planned.

When completed, the beach was about 125 feet wide. Dramatic *before* vs. *after* (*no-beach* vs. *wide-beach*) pictures (figs. 4.12a, 4.12b, 5.16, 5.17) affirm the immediate effectiveness of replenishment as a shore protection strategy. One must, however, go beyond the photographs taken on the day the dredge left town. A more important question is, "What did the artificial beach look like after three months? twelve months? two years?" The original plan was to maintain a beach 60 feet wide (later changed to 40 feet). Did we get our money's worth?

While the sand in the central part of the island stayed in place, and even built up impressive new dunes (fig. 5.12), the sand at the island's north and south ends quickly disappeared. The sand at the north end was to be anchored by a large $1 million groin the Corps built before they pumped the sand. However, this groin "leaked," allowing sand to flow over, through, and around it and around the northern shoulder of the island to the Savannah River shoreline (fig. 5.10). This groin, which was intended to last fifty years, needed $80,000 worth of modifications only four years after being built.

The sand at the south end of the island also disappeared at an unexpectedly rapid rate. To obtain the 2.3 million cubic yards of sand needed to create the artificial beach, the Corps dredged a tidal shoal approximately 2,000 feet from the south end of Tybee Beach in the inlet to Tybee Creek. Immediately following this dredging, accelerated erosion occurred on the island's south end. Only eight months later, waves were again crashing against the seawall. And just two years later, the south end parking lot had to be closed because of damage caused by wave action. During Hurricane Diana in 1984 waves were actually breaking onto the parking lot. This hurricane, plus a subsequent northeaster, placed a section of the seawall in danger of imminent collapse, and in 1985 emergency riprap reinforcement was brought in.

In 1987 the Corps undertook a new project

to (1) repair the leaky north groin, (2) build a new groin at the south end, and (3) pump up an all-new artificial beach (figs. 4.12a, 4.12b, 5.16, 5.17). The total cost came to at least $11.5 million. When the beach eroded at a rate faster than anticipated, the blame was placed on bad weather!

Inlet Migration. The Savannah River inlet is large and relatively stable, although the position of its MHW shoreline has historically undergone wide fluctuations, especially on the south side (northeastern Tybee shore). The migration of the main south channel and the associated flood tide channels are important causes of these shoreline changes. The construction of channel jetties has delineated the north channel of the Savannah River for some time now.

The creation—and subsequent erosion—of a large northerly spit marks that shoreline's most dramatic migration. Despite accretionary trends during the late 1800s, the northern inlet shore experienced dramatic erosion between 1924 and 1974 (fig. 5.9). A spit formed again in 1977–78 as sand from beach nourishment passed over and through the north end groin to accrete along the inlet.

5.16 South end of Tybee Island after the 1975–76 beach restoration, but before the 1987 replenishment. Early 1980s. Photo by W. Cleary.
5.17 South end of Tybee Island several months after the 1987 replenishment. February 1988. Photo by T. Clayton.

Dredging of the Savannah River channel has been implicated as a probable cause for some erosion on the inlet and ocean shores. Some geologists think the continuous deepening of the channel to the current 38 feet below mean water has facilitated removal of Tybee-bound sand from the nearshore sand-sharing system.

The Tybee Creek shoreline largely responds to changes in the inlet's ebb tidal shoals. This shore experienced net erosion between 1866 and 1974, except for a small section near the ocean beach. The entire inlet shoreline, however, experienced little or no accretion from 1925 to 1974 (fig. 5.9). Timber bulkheads along the inlet shoreline are evidence of past erosion problems (fig. 5.13).

According to a Corps of Engineers study of the south end ebb tidal shoals, a 20-year cycle of accretion and erosion exists at the juncture of ocean and inlet beaches. The cyclical migration of channels within the ebb tidal shoal system is thought to be responsible for these short-term patterns.

Storm Overwash. The potential for overwash is determined by storm conditions, upland elevation, maturity and height of sand dunes, and beach width and slope. Footpaths or roads that breach the foredunes and backdunes are likely to serve as paths for flooding waters. During the New Year's Day storm of 1987, large interdune areas were flooded with storm waters that breached these openings. Installing elevated overdune paths that cross both backdunes and foredunes will help to close any breaches in the dune system. Fortunately, most of the ocean beach areas with mature dune systems (Sixth to Thirteenth Streets) have crossings in place.

Areas with no substantial dune system are in greater danger of overwash/flooding during storms and occasionally during extremely high spring tides. In particular, the island's south end (from Fourteenth Street to the southern tip of the island) and the shoreline near Third Street are prone to flooding during periods of severe erosion. During the storm of January 1, 1987, ocean waters flooded 80–100 yards inland at the south end parking area and up to Butler Avenue near Third Street!

Storms. (See appendix B and the publications listed in appendix C for more detailed descriptions of hurricanes and storms that have struck the Georgia coast.)

In recent years several major northeasters have buffeted Tybee. Three times during the 1980s flooding occurred at the south end parking areas and some 50 to 100 yards inland on access roads (Sixteenth to Nineteenth Streets) as waves topped the seawall during periods of severe beach erosion (fig. 5.14a); several houses to the south of Eighteenth Street and some others in the vicinity of Third Street also suffered damage. Flooding of low-lying areas on the backside and river beaches of Tybee occurs during such storms as rising waters overfill the marshes.

Major hurricanes that affected the northern Georgia coast occurred toward the end of the nineteenth century. During the past century the "great hurricanes" of 1881, 1893, and 1898 were the most powerful and destructive to strike the region. The 1893 hurricane is classified as a 100-year storm, that is, one with a 1 percent chance of occurring within any given year. It inundated Tybee with a storm surge 19 feet above MSL. The combined effects of both tidal and wave action may have increased this surge to almost 25 feet.

Descriptions of the hurricane's aftermath indicate that inlets were created across Tybee, ships were stranded, and many structures were destroyed. In fact, the area between the beach and the railroad bed (near Butler Avenue) was described as "a desert having been swept almost completely clean of vegetation, structures, and sand dunes." This storm is estimated to have killed 1,000 to 2,000 people in the low-country region of Georgia and South Carolina.

Evacuation

Highway 80 (fig. 5.7) is the only access road for Tybee Island. Because of its low elevation and susceptibility to flooding, evacuation must be accomplished hours before the storm's actual arrival. For Georgia islands with causeways, emergency management officials anticipate that impassable floodwaters will halt evacuation at least five hours before hurricane landfall. High winds and heavy rains could force an even earlier cutoff. (The lowest point on Highway 80 is about 7 feet above MSL; the storm surge for the 1893 hurricane was 20 feet above MSL.) Low sections of the causeway are subject to flooding during spring tides as well.

Evacuations of the island are well-planned and effectively implemented. A specific evacuation plan for Tybee is being developed. Consult the Tybee City Hall and the Chatham County Emergency Management Agency (Civil Defense) for details.

When the order to evacuate comes, don't delay! *Do not* consider remaining on the island—during the hurricanes of 1881 and 1893, Tybee Island was completely inundated.

Services
Solid waste disposal, drinking water supplies, sewage treatment, and roads are addressed in the 1985 land-use plan. In 1979 the island landfills were closed to garbage disposal, and all garbage and trash are currently hauled off the island.

The city-operated waterworks, purchased in the 1950s and updated in 1974, provide drinking water to island residents. Major improvements in the entire delivery system were recommended by the study. A secondary treatment facility for sewage was built in 1975. The current system will allow for increased use until around 1995, but leaking, deteriorating lines are expected to hinder expansion.

Land Use
The establishment of subdivision plat maps in the late 1800s defined the eventual patterns of growth on Tybee. The current arrangement of the street system closely follows those early plans. The presence of the Fort Screven Military Reservation on the north end effectively delayed any development of residences in that area until 1946 when the land was opened for sale to the public. Until that time most growth had occurred on the island's south end. Even today, development over large portions of the north end is much less dense than in the south.

Savannah Beach—Plan for a Comprehensive Future was developed in 1975 as the first land-use plan for the island. In 1985 *A Comprehensive Plan for Tybee Island, Georgia, 1985–2005* was produced by a group of volunteer planners. This thorough, balanced study considers many facets of island life and community services.

Specific recommendations regarding land use are made for each "neighborhood" identified in the 1975 plan. The study addresses public opinion in these areas: future growth, anticipated population changes, the economy and local businesses (tourism-based), erosion control, parking and beach access, land-use plans and zoning ordinances, drainage patterns, and public services. In 1992 the 1985 plan was being updated; it will be reviewed and revised regularly. A copy of the plan may be viewed in the public library or at City Hall.

Little Tybee Island and Williamson Island

Tybee, Little Tybee, and Williamson Islands (fig. 5.9) are part of a single sand-sharing unit (fig. 3.12). Little Tybee and Williamson, however, have more in common with each other than with Tybee.

The Tybee-Little Tybee-Williamson Island complex is of Holocene age. The extensive marshes between the former barriers (now inland) of Pleistocene age and the present barrier complex of Tybee-Little Tybee-Williamson Islands are a product of sediment deposition within the Savannah River delta complex (fig. 3.6).

Little Tybee Island is actually an assemblage of over 40 beach ridges and hammocks (tree islands) surrounded by tidal marsh. The beach ridges make up only 600 of the island's 6,180 acres. The remainder is salt marsh or intertidal beach. Including the marsh, Little Tybee is about 4.0 miles wide.

A veneer of intertidal ocean beaches, 5.0 miles in length, fronts high marshes that in turn protect the hammocks of Little Tybee. These beaches are highly dynamic, and both Little Tybee and Williamson have unstable shorelines, sometimes migrating up to 183 feet in a single year!

Williamson, the youngest Georgia island, is about 1.5 miles long and less than 0.2 miles wide. Originally a sand spit at the south end of Little Tybee Island, it was disconnected by inlet breaching during the late 1950s (fig. 5.9). Occasionally, the sandy ridge is almost completely flooded at high tides. Varied dune vegetation has struggled to colonize and form some scattered dunes.

Little Tybee

Little Tybee Island hammocks (areas higher than their surroundings) have served as homesites for a handful of coastal residents. Native Americans probably used the area; colonists certainly lived on the hammocks. Nineteenth-century maps show homesteads on the hammocks of Little Tybee and adjacent Beach Hammock. These residents primarily used the area for subsistence agriculture and isolated homes.

The relatively wild nature of Little Tybee prevails today despite several proposed uses that would have dramatically changed its character. During 1938 the state, with the support of a citizen's group in Savannah, proposed developing a $600,000 state park. A bridge across Tybee Inlet, 150 cabins, bathhouses and boathouses, and a swimming pool were included in the plans.

The state park, however, did not materialize despite years of efforts. In 1955 the Laroche family sold the island to Franjo Inc. for $40,000. In 1957 the Georgia State Highway Department proposed constructing a bridge over Tybee Creek along with a causeway from Highway 80 and Lazaretto Creek to the island's largest hammock. A spur route to the beach was to be built after development of the hammock.

In 1962 Franjo Inc. filed a bold request for permission to build a causeway to Little Tybee from Tybee across Tybee Inlet! As noted by the *Savannah News-Press*, "the causeway would block off the Tybee Creek channel, causing a new beach to form on the east side of the proposed roadway. . . . Tybee Creek would be rerouted behind Little Tybee into Wassaw Sound." The controversy surrounding the original proposal then resurfaced as citizens demanded more work on existing roads to Tybee rather than any new road work; by 1965 plans for the causeway were dropped despite support from the governor and the highway department.

Franjo Inc. then announced plans for immediately developing Little Tybee into a privately owned beachfront resort with access only by water or air. Plans were made to rename the island but were eventually dropped.

Soon thereafter Little Tybee was purchased by the Kerr-McGee Corporation of Oklahoma. In 1968 the company applied for permission to strip-mine phosphate from deposits 40 feet below the marsh surface. Public protests over the proposal resulted in the state legislature passing the Marshlands Protection Act, which established legal safeguards for all tidal marshlands in Georgia.

In 1982 Little Tybee was incorporated into the federal Coastal Barrier Resources System. Being included in this system means that any construction or development is ineligible for federal subsidies (such as federal flood insurance, infrastructure assistance, or disaster aid).

In 1990 Kerr-McGee Corporation, the Nature Conservancy, and the State of Georgia reached an agreement that provides for the island's protection. Kerr-McGee donated Little Tybee Island, including Cabbage Island, to the Nature Conservancy. An anonymous gift of $1 million dollars enabled the state to purchase the island in 1991, and the Nature Conservancy used the proceeds to establish a tall grass prairie reserve in the Midwest. The state Department of Natural Resources will manage the islands as a wilderness preserve.

Williamson Island

Williamson Island is the youngest Georgia island. Shoreline surveys of 1913 show the formation of a sand spit from the central portion of Little Tybee's shoreline. At the time Beach Hammock was the southernmost ocean shoreline in the Tybee-Little Tybee complex. The spit continued to lengthen, with Little Tybee Creek routed to the south between the spit and the marsh. Between 1957 and 1960 the spit was breached. The resulting island is Williamson. This low, narrow barrier is like those typically found on coasts with high wave energy and low tidal energies.

After the island was created, Kerr-McGee Corporation and the state both claimed ownership. This disagreement over an ephemeral body of sand was taken to the courts. The state was eventually designated the island's proper owner and manager.

Wassaw Island

Wassaw Island (fig. 5.18), a 2.0-mile-wide Holocene beach ridge island (fig. 3.6), is the Georgia island least disturbed by human intervention. Only Fort Morgan (fig. 5.19) and a small housing compound have been built there since colonial times.

The 5.5-mile strand of oceanfront beach is characterized by an eroding shoreline along its northern third. Where erosion of forested

5.18 Historic net erosion and accretion: Wassaw Island, 1858–1974. (Adapted from Griffin and Henry, 1974.)

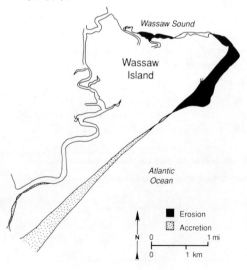

5.19 Fort Morgan, at north end of Wassaw Island, 1985. Photo by L. Taylor, courtesy of Georgia Marine Extension Service.

upland is occurring, the beach is littered with fallen trees. Sun-bleached skeletons of oak, pine, palmetto, and cedar lie in disarray along this expanse of the boneyard beach (fig. 3.16b).

The remainder of the ocean beach is backed by sand dunes. In particular, Wassaw's southeastern tip shows evidence of rapid accretion, with a series of multiple dune ridges parallel to the present shoreline. The remains of the Parsons family's south end beach pavilion, now one-quarter mile inland, can also be seen. Last used in the 1930s, the pavilion site is now surrounded by 40-foot pine trees!

This pattern of erosion on the north end and deposition on the south end is typical of several Georgia islands. As a result, the whole island has shifted, rotating counterclockwise, over the past century (fig. 5.18).

Limited future use of the island has been guaranteed. When the Parsons family sold most of the island to the Nature Conservancy, it was with the stipulation that bridges and camp fires be banned. The Nature Conservancy later deeded the island to the federal government (1969), and it is currently managed as wildlife habitat within the U.S. Wildlife Sanctuary system. The island is open only to daylight visitors; no development is allowed.

Ossabaw Island

Ossabaw Island (fig. 5.20), 9.1 miles long and 5.4 miles wide, is a Pleistocene island with a Holocene beach ridge fringe (fig. 3.6).

The Holocene portion of Ossabaw is characterized by former beach ridges separated by fingers of tidal salt marshes. Parallel rows of steep former dunes are covered with pines and mixed hardwoods. Rains during the wet season create seasonal ponds, or sloughs, between the former dune ridges.

The older, Pleistocene portion is distinctly different from the Holocene. Marshes—some tidal and brackish; others impounded and fresh—delineate the boundary between the two areas. Higher and flatter than the Holocene area, the Pleistocene portion of Ossabaw comprises much of the upland acreage. The vegetation of the maritime forest here is more diverse and mature than that of the Holocene strands. Permanent ponds in these forests provide freshwater habitat for fish, waterfowl, and alligators.

Studies of erosion and accretion on Ossabaw Island have noted the existence of three *nodal points* (fig. 5.20). Each marks the location of a relatively stable portion of shoreline; the areas to the north and south of each point exhibit large fluctuations in position, eroding and accreting. The net picture is one of accretion at both ends of the island and in the central section; shoreline recession has occurred in the segments in between. Overall, the island has grown since 1957, indicating that the Ogeechee River may be delivering some sediment.

The inlets at each end of the island are relatively large and stable. In fact, as mentioned, both sound shorelines have been accreting since 1957. (Before then, the Ossabaw Sound shore had been retreating.) However, the two small inlets associated with the tidal creeks that cut across the ocean beach are highly mobile. Washover fans have been mapped in the central and northern portions of Ossabaw where the dunes are relatively low (fig. 3.22).

Ossabaw Island is now a Georgia Heritage Wildlife Preserve (the first), managed by the Georgia Department of Natural Resources Fish and Game Commission. Visitation and development are strictly limited.

Ossabaw Island serves as an excellent example of how difficult it is to conserve natural resources under private ownership, even when the owner is deeply committed to preservation. For an interesting account of the struggles faced by Eleanor West, the former owner, in her effort to maintain Ossabaw in a relatively natural state, see Ann Simon's book, *The Thin Edge* (chapter 6: Litmus for the Coast).

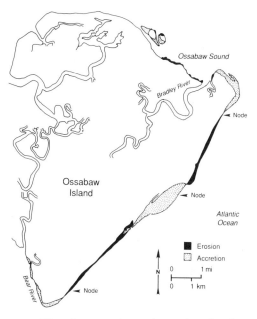

5.20 Historic net erosion and accretion: Ossabaw Island, 1924–1974. (Adapted from Griffin and Henry, 1984).

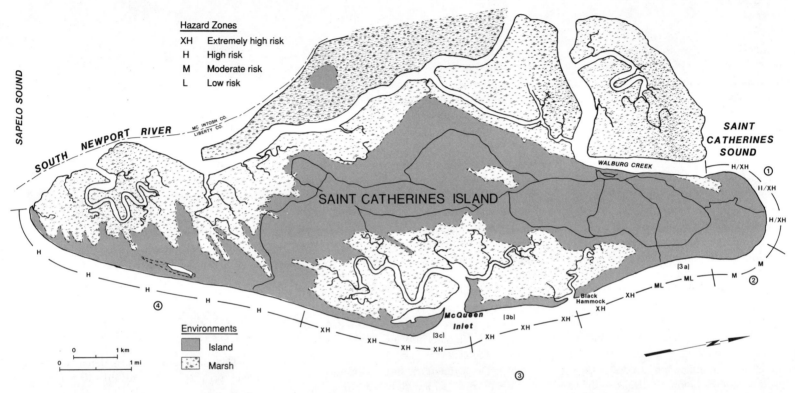

5.21 Site analysis map: St. Catherines Island. Companion map to table 5.7.

St. Catherines Island

St. Catherines Island (figs. 5.21 and 5.22; table 5.7) is another sea isle with a Pleistocene core and Holocene veneer (fig. 3.6). Its northern and central parts are Pleistocene components of a previous barrier shoreline. The northeastern, central, and southern beaches and ridges are Holocene in origin. The barrier beaches of the island's central section are separated from the Pleistocene core by tidal marshes that drain and fill through ocean-fed creeks (fig. 5.21).

One of the most striking scenes on the Georgia shoreline is the "sea bluff" of St. Catherines (fig. 5.23). Created by erosion of the high Pleistocene core, a steep 20-foot bank overlooks the beach near the north end. Marsh peat and upland soil outcroppings also are found along this stretch. This is the only such segment of Pleistocene shoreline along Georgia's ocean beaches.

St. Catherines is unusual among the Georgia islands in that it has experienced almost exclusively shoreline retreat over the past century (fig. 5.22), perhaps because of its great distance from any major river. Other characteristics of the island are summarized in figure 5.21 and table 5.7.

Designated as a National Historic Landmark site, St. Catherines is owned and managed by the St. Catherines Island Foundation; it serves as a research site for the American Museum, New York Zoological Society, and

84   Living with the Georgia Shore

Table 5.7 St. Catherines Island Shoreline Hazard-Risk Evaluation

Shoreline section	Comments
1	*Walburg Creek/St. Catherines Sound shoreline* Elevation of river bank averages approximately 10 feet; flooding potential relatively low Relatively severe, constant erosion since 1858 No known erosion control structures Steady southward inlet migration
2	*Northeast point to south end of dune line at sea bluff* Elevation of upland averages about 10 feet with former dune ridge nearshore ranging up to 20–25 feet; flooding not likely Accretionary trend since 1858 No erosion control structures Inlet migration at nearby St. Catherines Sound shoreline but with no recent adverse impact Dune overwash is infrequent; does occur during storms
3a	*Along sea bluff (from northern dune field to end of spit at Black Hammock and tidal inlet)* Elevation of island upland, 10–20 feet; salt marsh separates Pleistocene upland/island from Holocene beach ridge/spit at south end of section; beach ridge/spit elevation approximately 5 feet with 20-foot elevation on Black Hammock). Flooding potential of upland edges low to moderate; flooding potential of beach ridge relatively high Erosion rapid and constant along entire ocean beach since 1858; exposure of upland peat occurs on upper portion of section adjacent to upland No sign of erosion control structures Migration of tidal creek inlet at south end of spit has been cyclical; recent trend is erosional Dune overwash common in spit area; no dunes present to north of spit
3b	*Inlet at Black Hammock to McQueen Inlet* Elevation of upland (approximately 1 mile inland from beach and marsh) averages 10–15 feet; only upland/vegetated areas on beach strand are approximately 5 feet Flooding potential for upland low except on marsh edges; flooding on beach ridge may be relatively common Erosion constant and rapid since 1858 No sign of any engineering structures Rapid, constant inlet migration at Black Hammock Inlet and McQueen Inlet since 1858; highly dynamic Dune overwash of low-lying strand common during storm and spring tides
3c	*Spit/Strand south of McQueen Inlet to juncture with Pleistocene portion of island* Elevation of upland (approximately 1/4– 3/4 miles behind spit and marsh) averages 10–15 feet; elevation on beach strand, 5–15 feet Flooding potential on upland low to moderate, but strand probably flooded frequently No erosion control structures evident Active inlet migration only at McQueen Inlet Dune overwash highly likely during storms
4	*Holocene beach from Beach Hammock to south end of island* Elevation: upland adjacent to the beach, 5–10 feet Flooding potential moderate to high in low-lying areas adjacent to marsh and beach Shoreline changes erosional since 1858 with rapid and dramatic loss of sand on southeastern edge No evidence of engineering structures Inlet along Sapelo Sound shoreline has widened as shoreline has eroded Dune overwash likely during storms and high spring tides

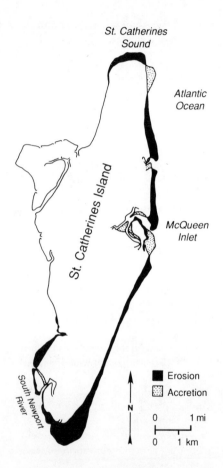

5.22 Historic net erosion and accretion: St. Catherines Island, 1958/67–1974. (Adapted from Griffin and Henry, 1984.)

many scientists. Due to the sensitive nature of the Wildlife Survival Center and associated research, the island is not currently open to visits by the general public or to extensive development.

Blackbeard Island

A Holocene beach ridge formation, Blackbeard Island is 6.4 miles long and about 2 miles wide (fig. 5.24). It is part of the Blackbeard-Sapelo (Cabretta-Nanny Goat) island unit (fig. 3.12). Its topography is characterized by numerous parallel dune ridges throughout the interior. Low areas between the ridges form sloughs, which flood during the wet season. Permanent ponds, large freshwater impoundments, and savannas and freshwater marshes are located throughout Blackbeard's northern and central portions. Communities of oak, pine, and some mixed hardwood make up most of the maritime forests. Shoreline changes here are dominated by two nodal points and unstable Cabretta Inlet (figs. 5.24 and 5.25).

Formerly a quarantine station for yellow fever victims, the island is now managed by the federal government as the Blackbeard Island National Wildlife Refuge. (The hospital for sufferers of yellow fever was located at the

5.23 The "sea bluff" of St. Catherines, 1981. Photo by L. Taylor.

south end of the island but was destroyed by the hurricane of October 1898). Although open for public day use, the island is currently off-limits to major development and probably will remain so. It has been incorporated into the federal Coastal Barriers Resources System.

Sapelo Island

The bulk of Sapelo Island (figs. 5.24, 5.25, and 5.26) is Pleistocene core, fronted by a strip of Holocene beach and dune sediments (fig. 3.6). A thin ribbon of marsh separates these two "generations" of barrier islands. With elevations of up to 23 feet, the Sapelo complex sports some of the highest ground on any Georgia island. Typical elevations on this wide, flat island are 10 to 15 feet.

Shoreline changes in the Cabretta Island (fig. 5.26) area are predominantly controlled by the cyclical migrations of the highly unstable Cabretta Inlet (fig. 5.25). Considerable erosion also has occurred at the island segment's south end. Two overwash fans have been mapped on Cabretta Island (fig. 3.22).

A nodal point exists at Nanny Goat Beach, with minimal net erosion to the north and considerable accretion to the south (figs. 5.24 and 5.26). Since 1924 erosion at northern Nanny Goat Beach has been steady.

Much of our knowledge of 19th-century Georgia hurricanes comes from Sapelo Island records. Thomas Spaulding's account of the 1824 hurricane tells of a "wall of water six feet

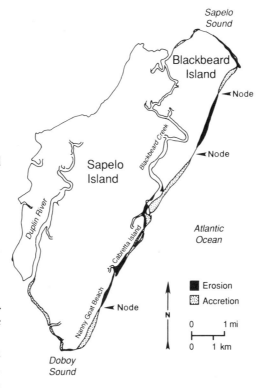

5.24 Historic net erosion and accretion: Blackbeard and Sapelo Islands, 1929–1974. (Adapted from Griffin and Henry, 1984.)

high [that] swept across much of the island." Several major storms during the 19th century flooded the whole of Sapelo.

The hurricane of October 1898 was severe. A storm surge of 12 to 18 feet inundated the island and caused substantial property damage. Archibald McKinley, a resident of southwest Sapelo, noted that "the waves coming across the island—direct from the ocean, covered the tops of our windows. In the house the water was nearly 3 feet deep—in the yard, nearly 6 feet deep on a level. The waves in our yard were fully 12 feet high."

Sapelo Island is currently owned by the state and is home to the Sapelo Island Research Foundation, the University of Georgia Marine Institute, the R. J. Reynolds Wildlife Refuge, the Sapelo Island National Estuarine Research

5.25 Cabretta Inlet is a highly unstable inlet which migrates rapidly to the south. (Adapted from Griffin and Henry, 1984.)

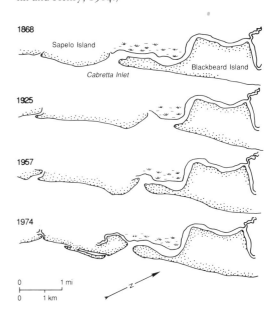

Selecting a Site on Georgia's Barrier Islands

5.26 North end of Cabretta Island, 1985. Photo by L. Taylor.

Reserve, and the community of Hog Hammock.

There is no bridge to the island, and it is slated to remain undeveloped, accessible only by boat. Public tours of Sapelo are offered through the Darien Chamber of Commerce by the Georgia Department of Natural Resources.

Wolf Island

Wolf Island (fig. 5.27) is a small Holocene island, 3.0 miles long and 2.7 miles wide. Maximum elevation is only 5 feet, even at the top of the highest dune ridge. The island is almost entirely salt marsh, with only a thin strip of beach sand that is being washed landward over the marsh. The oceanfront shoreline has shown continuous retreat since 1857 (fig. 5.27). The northeast end of the island has retreated 3,350 feet in that time.

This small washover island is owned by the federal government and is managed as a National Wildlife Refuge by the U.S. Fish and Wildlife Service. The island also enjoys further protection as a designated wilderness area. Development on the island, therefore, is not allowed.

Little St. Simons Island

Little St. Simons Island (figs. 5.28 and 5.29; table 5.8), the last of the family-owned barriers, is the northernmost of the barrier complex that includes Sea Island and St. Simons Island (fig. 3.12). Separated from Wolf Island to the north by the Altamaha Sound and from Sea Island to the south by the Hampton River, Little St. Simons is a complex of deltaic sediments associated with the Altamaha River. It is 5.2 miles long and 3.6 miles wide.

Elevation. Most of the island is marsh (fig. 5.29). Only about one-quarter of its total acreage is high ground, with most of the upland area a hammock surrounded by marsh. Elevation ranges from sea level to a maximum of 28 feet. The dune ridges are on average about 10 feet high.

5.27 Historic net erosion and accretion: Wolf Island, 1857–1974. (Adapted from Griffin and Henry, 1984.)

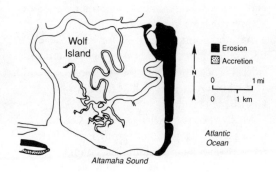

Historical MHW Shoreline Changes. Since 1866 the entire shoreline has accreted dramatically (fig. 5.28). This newly created land is low with scattered dunes, and the entire area is prone to flooding and washover. Structures built in the higher and more vegetated inland areas are at relatively low risk.

Although accretion is the general pattern, unstable areas of shoreline recession do exist. On barriers, areas of greatest instability often are shores adjacent to an inlet, which is the case for Little St. Simons. For example, the southern end has experienced some periods of recession (for example, between 1924 and 1974), apparently related to the Hampton River's migration. The northern end, along the Altamaha Sound, recently seems to have entered a new period of instability. Development along the river shoreline is at extremely high risk.

No known erosion control efforts have been tried at Little St. Simons.

Inlet Migration. The Altamaha River inlet at the island's north end and the Hampton River inlet at the south end are both relatively large and stable. Both inlet shores, however, exhibit some instability. Historically accretionary (1860–1974), the northern shore recently (1974–1982) has been receding. On the southern end, the reverse is true. The earlier erosional trend (1924–1974) recently (1974–1982) has been reversed. The small tidal creek that discharges across the ocean beach is quite mobile and would be a potential hazard to any structures in the area.

Storm Overwash. Most of the recently accreted land on Little St. Simons is low and prone to overwash. In addition, livestock grazing has prevented a primary dune system from forming over most of the island, thereby increasing susceptibility to washover. A primary dune system does exist on the island's southern one-third of the island.

Storms. Published storm accounts from Little St. Simons Island are rare because of its historically sparse population. However, the area was particularly hard hit by the second of the two hurricanes to strike Georgia in 1898. This storm buffeted the coast for a seemingly endless 18 hours and left in its wake some 80 dead on the coastal islands near the mouth of the Altamaha River.

Evacuation. Because there is no road access to Little St. Simons, visitors must evacuate especially early in the event of an approaching storm.

Land Use. Midden sites, some with pottery shards, indicate that Native American coastal dwellers used Little St. Simons. When English colonies became established, however, European settlers and rice plantations were introduced. In the early 1900s the island was sold to the Berolzheimer family for its cedar trees. Their intent was to turn them into pencils; however, the gnarled wood of these small, twisted, coastal trees proved unsuitable. The island is currently managed by the family as a small, exclusive nature resort; only a small area on the backside has been developed.

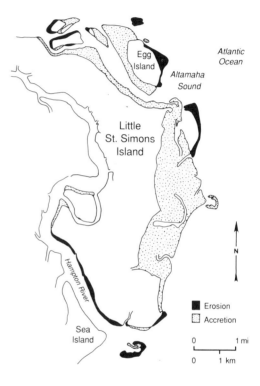

5.28 Historic net erosion and accretion: Little St. Simons Island, 1860–1974. (Adapted from Griffin and Henry, 1984).

Much of the rest of the island has been open to free-ranging livestock (horses, cattle, and European fallow deer). These animals have had a significant impact on the vegetation and therefore the topography of the island; in particular, their grazing prevented a primary dune

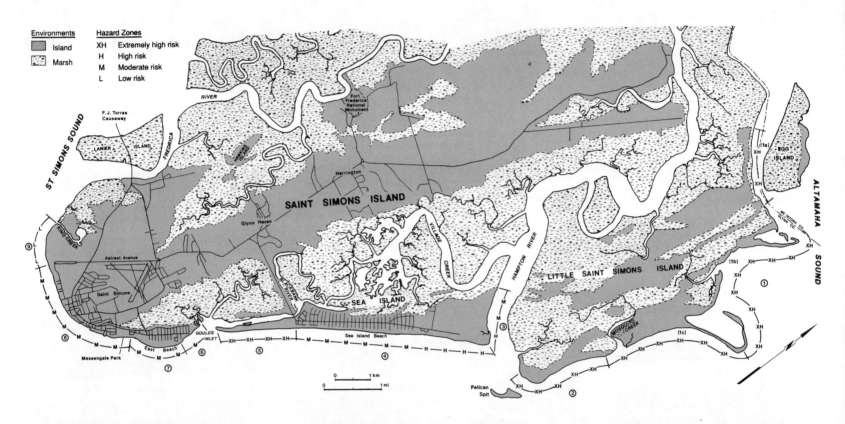

system from forming along most of the ocean shore. More recently, the cattle have been removed, and the deer population has been reduced. New dune and forest vegetation can now be established.

This island in 1982 was among the first to be incorporated into the federal Coastal Barrier Resources Act system. As a result, any development is ineligible for federal subsidization (for roads or federal flood insurance, for example).

Sea Island

Sea Island (figs. 5.29, 5.30a, 5.30b; tables 5.3 and 5.9) is the state's southernmost Holocene island, 5.1 miles long and 2.1 miles wide. It has slightly more marsh than upland area.

5.29 Site analysis map: Little St. Simons, Sea, and St. Simons Islands. Companion map to tables 5.8, 5.9, and 5.10.

Native American coastal dwellers used Sea Island; however, they, like the early European explorers, seemed to prefer adjacent St. Simons. Sea Island, which has been called Goat Island and Long Island, has served as light-

Table 5.8 Little St. Simons Shoreline Hazard-Risk Evaluation

Shore-line section	Comments
1a	*Low sand/shell ridge along Altamaha River shoreline* Elevation of ridges generally low enough to be flooded on high spring tide; backed by marsh Accretion along shoreline has been continuous since 1860, averaging 6.6 feet per year (with gains of over 21 feet per year, 1954–1974) No engineering structures Inlet relatively stable No dunes present; storm overwash of ridges
1b	*North-central shore* Narrow sand ridge with dunes approximately 6 feet high backed by high salt marsh; recent accretion Accretion constant since 1860, with maximum gain of 5,170 feet 1924–1974; recent (1974–1982) slowing of accretion No engineering structures Adjacent to northward-migrating inlet Dune system developing; most dunes low
1c	*South-central shore* Generally, low-elevation dunes (<8 feet) backed by high marsh Continuous accretion since 1860, with maximum gains of 70 feet per year, 1974–1986; 1924–1974, slowdown to 25 feet per year No engineering structures Migrating creek inlet (to north since 1860) at south end of section Overwash of dunes possible in large storm
2	*Mosquito Creek/Hampton River* Forested beach ridges Upland elevation to 10 feet; dunes to 15 feet Net accretion since 1860 along most of ocean beach; Hampton River shoreline erosional, 1924–1974; however, area accreted up to 380 feet, 1974–1982 No engineering structures Migrating creek inlets affect entire section Overwash of dunes extremely infrequent

Table 5.9 Sea Island Shoreline Hazard-Risk Evaluation

Shore-line section	Comments
3	*Hampton River shoreline* Upland elevation ranges from 5-foot to 10-foot rises; salt marsh strip separates upland and river Recent accretion, 1924–1974; long-term (1860–1974) history of erosion No erosion control structures Shoreline sensitive to movement of Hampton River channel; recent erosion (since 1980) on river shore
4	*Oceanfront shoreline: Hampton River to north end of spit* Upland elevation approximately 7–10 feet; flooding/overwash associated with storm activity Historical trend of erosion along entire strand Almost entire length of section lined with granite revetment or concrete seawall; some groins Inlet affects shoreline at Hampton River entrance; recent trend of erosion Overwash of revetment occurs during major storms; no continuous dunes remain Recently, movable groins and replenishment have been used
5	*Oceanfront Shoreline: Sea Island spit* Elevations on this dune-covered spit range up to 15–20 feet Rapid elongation of spit (5,840 feet) 1857–1974; some recent erosion No engineering structures Southward migration of Goulds Inlet has accompanied rapid spit growth Overwash of dunes occurs most frequently at southern tip of spit

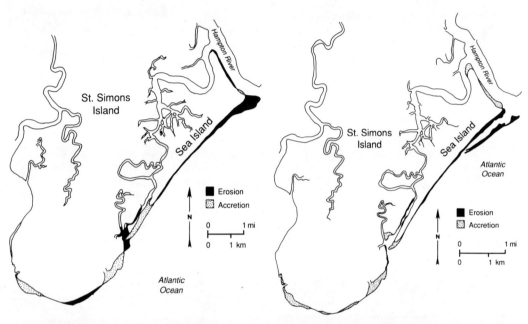

5.30 Historic net erosion and accretion: Sea Island and St. Simons Island. (Adapted from Griffin and Henry, 1984). (a) 1857/60–1974. (b) 1924–1974.

house site, plantation lands, grazing area, and recreational playground.

First opened to easy access in 1924 when bridges and causeways to Sea and St. Simons Islands were constructed, Sea Island has grown into a quiet, manicured, exclusive resort and residential community. Howard Coffin, who owned the island in the mid-1920s, began developing the resort; it is now managed by the Sea Island Company.

Development is currently concentrated in the central portion. Island developers have avoided building on much of the Hampton River shore and on the narrow south end spit. Much of Sea Island's northern end is undeveloped.

Because of the island's location on a river shore, any development on the northern shorefront will be at high risk. Low-risk sites are available in the high-elevation areas of the forest.

The oceanfront homes are sited extremely close to a chronically receding shoreline, which has resulted in the construction of concrete and stone walls along the entire length of the developed oceanfront and in the loss of the natural beach. The Sea Island Company has pumped up sand in some oceanfront areas and would like to establish a dune system here.

The island's oceanfront has a serious problem with historical erosion; now protected by a revetment wall and a nourished beach, development along the central oceanfront is classified at moderate risk. Low-risk sites are available inland. Check a topographic map available from the U.S. Geological Survey (listed in appendix C) to identify areas of highest elevation.

The Goulds Inlet spit is currently undeveloped. As such, it was incorporated into the federal Coastal Barrier Resources Act system in 1982. Development on this land is ineligible for federal financing.

The Goulds Inlet spit at the island's south end has lengthened 5,840 feet from just 1857 to 1974 (figs. 5.30a, 5.30b). The spit, as you can see, has undergone considerable historical change. It is highly dynamic and unstable.

Elevation. Elevations on the upland range from sea level to 16 feet; typical elevations are about 5 to 10 feet. The highest elevations (more than 20 feet) are on the dune-covered south spit.

History of MHW Shoreline Changes. Sea Island has experienced net erosion along its entire length since 1860 (fig. 5.30a). The history of change has two phases: from 1860 to 1924 the Hampton River shoreline retreated up to 320 feet, while the oceanfront area accreted up to 250 feet. Since 1924 this pattern has reversed: the Hampton River shoreline has since accreted and the oceanfront has receded (fig. 5.30b).

Hampton River Shore. The earliest records of shoreline change here show moderate erosion (about 5 feet each year). Since 1924, in contrast, the section has been accretional (fig. 5.30b). This accreted land should not be considered permanent. From 1974 to 1980 the trend was reversed, and the shoreline receded at an average rate of about 9 feet a year. Changes on this section of shore are closely tied to the Hampton River channel's condition and position.

The island's northeast tip has eroded considerably and continuously since 1857/60 (fig. 5.30a). Future shoreline changes in this area may be affected by any offshore shoal dredging undertaken for nourishment or another purpose.

Northeast Point to North End of Spit. This section of open ocean shore was stable or even accretionary between 1860 and 1924. Since 1924, however, the story has been one of continuous recession—up to 380 feet between 1924 and 1974 (fig. 5.30b).

Shoreline changes between 1978 and 1988 indicate that these trends have continued. The northeast beaches (northern mile of ocean shoreline) eroded at a rate of 6 to 10 feet per year. The north-central beaches (between 28th and 36th Streets) were slightly erosional. The central and south beaches, however, experienced erosion of 10 to 20 feet per year. According to a 1988 consultants' report, wave

5.31 Sea Island sloping concrete seawall, 1981. Photo by L. Taylor.

refraction studies indicate that the area with the highest erosion potential is between the north end (at the Hampton River inlet) and the south end Beach Club. Historical shoreline change patterns support this conclusion.

Almost this entire shore was walled during the last two decades (figs. 5.31, 5.32), essentially halting the shoreline's recession. The 1990 beach replenishment provided a protective and recreational beach. Be aware,

5.32 Sea Island revetment, 1981. Photo by W. Neal.

however, that the processes responsible for the chronic erosion continue to affect the shore.

Spit Shoreline. Changes at the south end of Sea Island have been controlled by the rapid southward migration of Goulds Inlet (fig. 3.18). The result is 5,840 feet of spit elongation between 1857 and 1974 (fig. 5.30a), sometimes at a rate of up to 50 feet per year.

The spit's ocean shore has shown minor recession in that time. An incomplete 1980 aerial survey revealed an increase in recent erosion. Between 1978 and 1988 the south end of the spit continued to accrete, while the remainder was relatively stable.

History of Erosion Control Efforts. Prior to 1979, only the Cloister and a few private homes had seawalls or revetments. Since 1981, however, nearly the entire ocean-facing shoreline, with the exception of the Goulds Inlet spit, has been reinforced with seawalls of various shapes and designs (figs. 5.31, 5.32). Even property with no beachfront structures (the northern picnic area, for example) has been included in this mass stabilization.

The earliest history of stabilization attempts began much earlier. The Brunswick-St. Simons Causeway Bridge was built in 1924, opening St. Simons and Sea Islands to easy automobile access. Subsequently, Sea Island began to be developed in earnest (1926). The Sea Island Company early on built a revetment near the Beach Club. Hurricane Dora (1964; category 2) destroyed this revetment, which was replaced by another at North First Street. Also in response to Hurricane Dora, the Corps of Engineers pumped up a 1965 protective beach (150,000 cubic yards) along 4,000 feet between 25th Street and 36th Street. (One has to wonder at spending $200,000 of public money to protect exclusive, private development.) The Sea Island Company funded similar replenishment near the Beach Club between 1964 and 1968.

Between 1972 and 1981 almost the entire developed beachfront was revetted. Hurricane David (category 2) in 1979 spurred what seemed like an almost-overnight armoring of the island. Then, in 1981 high spring tides coincident with a passing northeaster destroyed the concrete seawall/revetment between 25th Street and 36th Street (fig. 3.21). In response, this wall was rebuilt and extended an additional 3,900 feet north of 36th Street. More recently the revetment was upgraded to a cement retaining wall. Quite predictably, this has resulted in the disappearance of the recreational and protective beach.

By the mid-1980s the condition of the beach

at the Beach Club had deteriorated so that residents proposed to pump sand from the Blackbank River and hold it there with a new 450-foot groin. Residents of downdrift East Beach, however, were concerned about the groin's effects on their property. The state also initially recommended against building the groin, preferring more frequent beach renourishment. However, permits were eventually granted for the sand-pumping and groins (with removable slats), and the project was constructed in 1986.

In 1989 the Sea Island Company applied for permits to pump 2 million cubic yards of sand from a Hampton River delta shoal to the beach between 2nd and 21st Streets; this area was replenished in 1990. A dune restoration project has been planned in conjunction with the replenishment; terminal groins are to be installed at the island's north end and to the south of the Beach Club a year or more after the sand-pumping. Be aware that replenished beaches must be continuously maintained and that these costs must be borne by island residents and developers so long as there is no public beach access.

The Goulds Inlet spit remains the only extensive unwalled portion of the island oceanfront.

Inlet Migration. The Hampton River inlet is relatively large and comparatively stable. It has shifted slightly to the south in the past century or so.

Goulds Inlet is highly unstable, historically migrating rapidly to the south. It is likely that the narrow Sea Island spit will be breached in the future to begin the migratory cycle over again. Such an event has been documented at Cabretta Inlet to the north (fig. 5.25).

Storm Overwash. A few low-lying areas are subject to overwash during high tides and storms. These include (1) the tip of the south end spit and (2) areas of erosion along the shores of the Hampton River and inlet. The upland elevation is generally high enough to prevent overwash during most storms, provided a sufficiently wide beach is present. However, when spring high tides coincide with significant storms, water may wash over areas that are rarely affected.

Before the 1990 beach restoration no high-tide beach existed in many areas of this replenished stretch. The MHW line was near the top of the rock revetment, the crest of which was regularly topped in several areas when northeasters struck coincident with unusually high tides. Waves washed seawater into the yards of some residents, and some overwash extended farther inland along beach access roads. Since Sea Island has pumped up a replenished beach, the threat of overwash will be diminished so long as that beach remains.

Storms and Hurricanes. Sea Island's narrow width, as well as the relatively low elevation of much of its land, makes it susceptible to flooding in a substantial hurricane. The storms of 1804, 1824, the 1890s, and 1964 all caused significant damage.

Northeasters have done considerable damage to Sea Island beachfront property. Major storms during the 1890s destroyed concrete seawalls and revetments (fig. 3.21). Flooding of upland property and some streets also occurred.

Evacuation. The Sea Island causeway is the island's only vehicular egress point. It runs for about a mile between St. Simons and Sea Islands and crosses a bridge over Blackbank Creek. The bridge is elevated slightly higher than the causeway, and this route may be subject to inundation during a hurricane's approach. Sea Island's limited population should allow speedy evacuation.

More important, however, the Sea Island evacuation route leads directly to the more heavily populated island of St. Simons! The route continues across St. Simons Island and to the mainland via the recently rebuilt F. J. Torras Causeway with two new high bridges. So, while Sea Island residents may have no trouble getting off their island, they may be forced to wait with thousands of other evacuees who will be leaving St. Simons.

Evacuation route maps are published in your local phone book, and a copy of the CGRDC general plan for evacuation of Glynn County areas is available at the CGRDC office in Brunswick.

Services. As part of the St. Simons Island Urban Service District, Sea Island's water utilities are provided by the Glynn County Water and Sewer Department. Drinking water for the island is pumped from the Floridan aquifer at a well on St. Simons Island.

The island is not serviced by the county sewage treatment system. Some houses still have septic tanks, while other areas use a private sewage treatment system.

Land Use. Sea Island is privately owned, and there is no public beach access. The island's northern end remains undeveloped. The island's central portion has mostly single-family residences, a few small condominiums, and the famous Cloister beach club and resort. The southern portion remains undeveloped and is currently used for general recreation, including nature walks led by local naturalists.

The 1990 *Glynn County Comprehensive Plan*, prepared by the Glynn County Department of Community Development, incorporates portions of an earlier (1981) report, *Comprehensive Plan for St. Simons and Sea Islands*, and provides a blueprint for management of the county's natural, historical, cultural, and sociological resources through the year 2010.

The Future Land Use Map for the Sea Island area shows that low-density urban use is planned for the northern 80 percent of the island, with tourist facilities on the remainder (near the Cloister).

St. Simons Island

St. Simons Island, 11.6 miles long and 3.8 miles wide with an area of more than 36 square miles, is one of the larger Georgia islands (figs. 5.29, 5.30a, 5.30b; tables 5.2 and 5.10). However, because much of this Pleistocene barrier lies behind its Holocene "partners" (Little St. Simons and Sea Islands), it has only 3 miles of open-ocean shore (fig. 3.6). The Frederica and MacKay Rivers and associated marshes (up to 5 miles wide) separate the island from the mainland and the city of Brunswick.

St. Simons Island was opened to development in 1924 by construction of the Brunswick–St. Simons Causeway Bridge, but the principal period of intensive development did not begin until about 1940. Today, it has grown into a popular resort and suburb-style residential area. Twelve square miles are of relatively high elevation, making much of the island relatively safe for development.

The northernmost St. Simons beachfront is in the Goulds Inlet area (fig. 5.33). Beachfront houses here are at moderate to high risk because of the absence of a beach and the inlet's well-documented southerly migration (fig. 3.18). Failure of the seawall would place this area in the extremely high risk category. A mid-1980s study by the Corps of Engineers indicated that the present revetment is sinking and may not be functional beyond the mid-1990s. Breaching of the Gould Inlet spit, however, would relocate the threatening inlet and bring much sand to the area.

Continuing south along the beach to the King-and-Prince Hotel, a well-known landmark, the beach widens considerably (figs. 5.34, 5.35), up to 300 feet in some places. Note how securely set back the first row of houses is. Such has not always been the case. Old-time East Beach residents tell stories of ocean waves knocking at their windows. This recent accretion (figs. 5.30a, 5.30b) is likely linked to the present position of Goulds Inlet; if the inlet is relocated to the north by a storm, some of the accreted land may disappear, as it has at Cabretta Inlet. Shorefront structures would then be at greater risk.

Massengale Park is the site of a former inlet, now closed.

The circumstances of the accreted bulge of sand at East Beach illustrate the problems that rise when shifting sands are regarded as real estate. With today's coastal land values, you can be sure that whenever a new foot of land appears above the high-tide line, someone is waiting to claim and develop it. At East Beach the state maintained that the people of Georgia were bequeathed this (perhaps temporary) gift from nature. Local developers, however, viewed the situation differently, and a lawsuit over construction of a hotel on the accreted sand resulted.

At the King-and-Prince Hotel the nature of the shoreline significantly changes. The section between the King-and-Prince and the St. Simons Lighthouse has a history of moderate to severe erosion. Although minor accretion near the King-and-Prince began in the 1960s, the island's southeastern tip receded 600 feet between 1860 and 1974.

Table 5.10 St. Simons Island Shoreline Hazard-Risk Evaluation

Shore-line section	Comments
6	*Goulds Inlet shoreline (Ocean Road to Twelfth Street)* Elevation along inlet shoreline approximately 5 feet; dunes close to ocean approximately 10 feet; flooding over revetment possible during storms Long-term, steady retreat of shore, and spit growth on Sea Island before inlet shore was "stabilized"; oceanfront south of inlet shore accreted Revetment lines inlet shore; groins have been built in past, as has seawall Rapid southward migration of Goulds Inlet Overwash/flooding of revetment, dune overwash common during storms
7	*East Beach area (Twelfth Street to Massengale Park area)* Upland elevation 8–10 feet; highest dunes approximately 10–12 feet; flooding of dunes may occur during storms Net accretion since 1860; recent erosional trend No engineering structures in place or used in recent years Southward migration of Goulds Inlet responsible for accretion Dune overwash accompanies storms and extremely high spring tides (infrequent)
8	*Massengale Park to Retreat Avenue* Upland elevation: 5 to 11 feet; discontinuous dunes on northern portion approximately 5 feet; flooding likely among dunes during storms and along edge of remaining armored shoreline Although net erosion has occurred along most of shoreline since 1860, stability or accretion has occurred since 1924 Granite revetment lines entire stretch of shoreline Migration of inlet channels and channel dredging may be linked to erosion losses Note: Park is site of former inlet; closed in 1955
9	*Retreat Avenue to King Creek* Elevation of upland ranges from intratidal marsh to approximately 10 feet; no dune system; flooding of adjacent uplands common during storms Historical trend of accretion since 1857; approximately 1,060 feet of accretion, 1857–1974 No major engineering structures Shoreline changes may be linked to channel dredging

From this area south all the way past the public pier on St. Simons, the recreational beach has been severely damaged because of the rock rubble seawall (figs. 5.36a, 5.36b). As early as 1968, much of this section had no beach at high tide.

Oceanfront development behind the seawall should be considered at moderate risk. The area is one of historical erosion, the protective beach has been destroyed, and ocean waves can top the seawall. These hazards are mitigated, however, by the revetment, which now defines the shoreline's position.

Soundside development behind this seawall is at lower risk, primarily because the sound is protected from direct ocean wave attack. Inlet shores, however, are subject to drastic and rapid changes. The area between the St. Simons Lighthouse and a point 3,000 feet to the west has experienced consistent erosion—as much as 800 feet since 1870. The lack of a protective beach, a history of erosion, and proximity to the inlet put shorefront property in this area at a moderate risk.

The section west to King Creek (5,200 feet) has been accreting, at least since the mid-1800s. Protected from open ocean swell, this area is relatively safe for development, provided the recent history of accretion is not reversed.

Elevation. Island elevation ranges from sea level to 21 feet at the top of the highest interior ridges. Most of the island is 10 to 15 feet above sea level. Only a few primary dunes

5.33 Goulds Inlet area of St. Simons Island. Early 1980s. Photo by W. Cleary.

remain; they are found in the southwestern portion of the island.

Historical MHW Changes. Major overall changes on St. Simons Island from 1857 to 1974 include (1) the rapid southward migration of Goulds Inlet and (2) a landward shift of the island's southern tip.

Goulds Inlet Shore. In historical times the southern shore of Goulds Inlet, in conjunction with lengthening of the Sea Island spit (fig. 3.18), has moved steadily to the south. Any site immediately downdrift of a migrating inlet is at extremely high risk. In this case, however, stabilization of the southern inlet shore has mitigated this hazard. Risk will be further reduced if a storm someday relocates Goulds Inlet to the north.

East Beach Area. This area (south of inlet shores to the vicinity of Massengale Park) has accreted about 1,150 feet since 1857/60. This accretion is associated with the closing of a small inlet at Massengale Park (between 1924 and 1955) and the southward migration of Goulds Inlet.

Southeastern and St. Simons Sound Shores. The southern tip of St. Simons Island has shifted landward (1857/60–1974) through erosion on the southeast shore and accretion on the southwest (fig. 5.30a). This shift may partly result from dredging of the Brunswick Harbor entrance channel. Future shoreline changes in the southeastern area are likely to be relatively small, since the shoreline here is now entirely stabilized by riprap and revetment.

History of Erosion Control Efforts. At the island's northern end (fig. 5.33), migration of Goulds Inlet (fig. 3.18) has prompted homeowners to build wooden bulkheads and dump rock rubble wall along the shore. Most of these are light-duty structures, however, and offer little protection from wave damage.

Somewhat more substantial protection is offered by the 3,000-foot rubble mound seawall built by the Corps of Engineers in 1964 in response to Hurricane Dora (Goulds Inlet to 10th Street). Protection offered by this seawall also is limited, however; it is designed to protect against a 10-year storm surge, and parts of the wall had already begun to fail by 1968, only four years after construction. And, as noted, the wall is sinking and may not function as intended beyond the mid-1990s. As expected, today there is no beach near Goulds Inlet.

Early attempts to stabilize the southeastern stretch of shore consisted of stone and rubble mounds placed on the beach by various homeowners. In response to Hurricane Dora in 1964 the Federal Office of Emergency Planning of the Corps of Engineers built the 8,000-foot-long granite and rubble revetment that stretches from the King-and-Prince Hotel to a point 1,900 feet west of the municipal pier. The structure is designed to protect against a 10-year storm surge. Once again, public money ($500,000 in this case) was expended to protect private property, and the result was destruction of the public's beach (figs. 5.36a, 5.36b).

Federal beach erosion control studies for this area were conducted in 1927, 1939–40, and 1968. No actions were recommended in any of these studies. The *Feasibility Study of Glynn County Beach Restoration* (1988) suggests that the following actions be considered by the county for St. Simons's beaches. A 2,650-foot beach replenishment from Goulds Inlet to 10th Street at East Beach was considered but not strongly supported. As an alternative to renourishment, construction of a revetment along 2,150 feet of shoreline was suggested. Relocation of Goulds Inlet up to one mile north of the present point also was considered.

Along the southern shoreline, from First Street at East Beach to a point immediately

north of the pier, 9,200 feet of beach replenishment was proposed. The placement of two to three groins was also suggested for the southern end of this project area.

Based on this 1988 study, the Savannah District office of the Corps prepared a proposal for a onetime replenishment. In early 1991 the plan was under review by the Washington office of the Corps. If approved by that office, the plan will be initiated as soon as Glynn County allocates $1.7 million to supplement $3.1 million in federal funds. Revenue bonds to be issued by the Glynn County Economic Development Authority are expected to provide the needed funds.

The planned replenishment will place sediment along two miles of beachfront on the island's south end. This fill will be composed of sediments removed from the Brunswick navigation channel during maintenance dredging, which will widen the channel by 100 feet. The renourishment, however, is only a onetime event, linked to the planned dredging of the channel bar. A proposal for initiating a long-term (probably 50 years) project has been proposed by the Corps' office. This project probably would require 51 percent federal and 49 percent state/local support. For a beach to be maintained seaward of the present line of

5.34 Aerial view of East Beach area of St. Simons Island. Goulds Inlet is visible to the right. 1987. Photo by L. Taylor.

5.35 East Beach. 1985. Photo by T. Clayton.

Selecting a Site on Georgia's Barrier Islands 99

riprap, an ongoing, long-term project is required.

Inlet Migration. St. Simons Island has been greatly affected by the rapid southward migration of Goulds Inlet (fig. 3.18); 1,640 feet of shoreline retreat occurred between 1857/60 and 1974.

Unlike Goulds or Cabretta, the inlet at St. Simons Sound is not an actively migrating inlet, but its shores are not static. Since 1857/60 the inlet has widened, and the southeast shore of St. Simons has retreated, while the southwest shore has accreted. The net result is a landward shift of the south end.

Storm Overwash. Some East Beach foredunes have been breached in winter storms (for example, the winter of 1987). Overwash potential is greatest at breaks in the dune line and in areas of erosion. Much of the St. Simons revetment is susceptible to overwash during storms.

Storms. Since 1893 the St. Simons area has been struck by five major hurricanes (averaging about one every 20 years).

Aaron Burr, who experienced the 1804 hurricane at St. Simons, wrote that the waters

5.36 (a) St. Simons Island south shore in the 1930s. Note the wide beach on which cars are parked. Photo courtesy of Coastal Georgia Historical Society. (b) That same shore, roughly half a century later. View is from the pier visible in fig. 5.36a, toward the lighthouse. No beach! Early 1980s. Photo by W. Cleary.

100 Living with the Georgia Shore

of the 7-foot storm surge were "sufficient to inundate a great part of the coast, to destroy all the rice, to carry off most of the buildings that were in the lowlands and destroy the lives of many. . . ."

More information about St. Simons hurricanes is given in appendix B.

Evacuation. Only one highway, the F. J. Torras Causeway, provides vehicular access to St. Simons. The recent widening of the entire causeway to four lanes and the opening of two high-span bridges will allow safer and faster evacuation. The causeway's elevation also was raised—an important modification to prevent early flooding of the road. Because this is the only road to the island, the early evacuation of St. Simons is important. The island's large population, plus the residents of Sea Island, may create heavy traffic flows. And once on the mainland, island residents must yet wind their way through the low-lying Brunswick area!

Services. Drinking water for the island's northern end and most of the southern half is provided by the Glynn County Water and Sewage Department through six wells that pump water from the Floridan Aquifer. Several residential areas, however, are not served by this system. Water pressure in many of these areas is inadequate to provide a strong flow for fighting fires. Residents should be aware of the capabilities of their systems. Five water storage facilities are located on the island within the county's water system.

Two wastewater treatment facilities service the island's southern half and northern tip. Wastes undergo secondary treatment (through extended aeration); effluent is disposed to Dunbar Creek. The system operates at near capacity during heavy rains, but average flows are well within the capacity of the plants. Still, many areas are not serviced by this system. Private package treatment systems and/or septic tanks serve many neighborhoods.

Land Use. Historical accident and the Future Land Use Plan element of the 1981 *Comprehensive Plan for St. Simons and Sea Islands* have so far guided patterns of growth on St. Simons. The majority of the population now lives on the island's southern half. As population increases, continued development of the northern portions will occur. (The Sea Island Company owns an estimated 95 percent of undeveloped land on St. Simons.) If the Glynn County Comprehensive Plan is followed, St. Simons will be densely populated when fully developed—in fact, the most populous island on the Georgia coast.

Past, present, and future land uses are discussed in (1) the 1981 *Comprehensive Plan for St. Simons and Sea Islands*, and (2) the 1990 *Comprehensive Plan for Glynn County*.

Jekyll Island

Jekyll Island (figs. 5.37 and 5.38; tables 5.4 and 5.11), the smallest of the sea islands, is another Pleistocene core with Holocene beach ridges in the north and south areas (fig. 3.6). It is approximately 7.4 miles long and 2.3 miles wide and is separated from the mainland by 4.0 miles of marsh. Of its 5,700 acres, about 1,400 are marsh.

The island's interior is characterized by concentrations of development, which are bordered by maritime forest and wetlands. Despite the modification of some areas, these forests remain relatively wild. Live oaks, lianas, epiphytes, palmettos, mixed hardwoods, and pines can be found.

Freshwater wetlands throughout the island act as seasonal sloughs and permanent or temporary ponds. Nestled between former dune ridges or within the heart of the forest's lowlands, these small and scattered wetlands provide critical habitat for many species.

In 1947 the island was purchased from the Jekyll Club by the state to serve as a state recreation area. In 1954 a bridge and causeway were constructed to link the island with the mainland, and since then considerable development has occurred.

Like Tybee Island, Jekyll Island may have been affected by navigation dredging in adjacent waters. Erosion on Jekyll increased significantly after dredging of the northern shipping channel was begun in 1909. Sand is no longer naturally bypassed across the inlet by the inlet shoal transport system. Instead, the sand is trapped in the deep channel and is eventually dredged up and dumped offshore.

With the important exception of the beach-

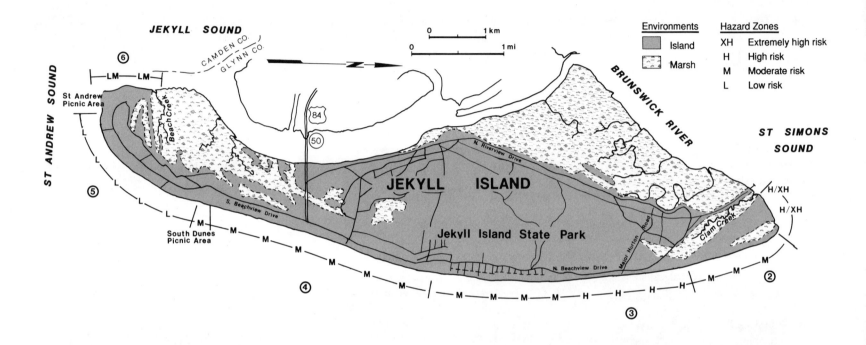

5.37 Site analysis map: Jekyll Island. Companion map to table 5.11.

front, development is at low risk on Jekyll Island. However, shoreline erosion and localized flooding during storms are of concern.

Well-protected from open-ocean storms, the island's western edge is the most sensible place to build. In fact, the island's oldest remaining residences are located along its west side. Although this area is not subject to the dramatic changes of the ocean beach area, changes along the banks of the backside rivers and channels do sometimes catch unwary homeowners by surprise. In the vicinity of the Jekyll Club Village a bulkhead has been constructed to prevent further upland erosion. Pay attention to elevation to avoid inundation by storm floods. The best sites are those somewhat inland with high elevation, set well back from riverbanks and marsh fringes.

Because of the general retreat of the northwest and north shorelines, no dunes exist in this area. At the north end, erosion has been continuous, with 900 feet of recession measured between 1856 and 1967, producing an average annual rate of more than 8 feet. Erosion continues to be rapid, evidenced by the toppling of trees growing on old beach ridges. This zone is at high risk for development.

Along the central portion of Jekyll Island, erosion has progressed at a slow but measurable rate. Many structures were constructed close to the beach in years past and are now

Table 5.11   Jekyll Island Shoreline Hazard-Risk Evaluation

Shoreline section	Comments
1	*Brunswick River shore (in St. Simons Sound)* Elevation of upland adjacent to shoreline approximately 5 feet, with a few 10-foot and 15-foot ridges; flooding of low-lying land adjacent to marsh probable during storms Erosion has continued at steady, accelerating pace since 1857 (1857–1924, 7.8 feet per year; 1924–1957, 11.5 feet per year; 1957–1974, 15.6 feet per year; 1974–1980, 5–10 feet per year) Shoreline armored with granite revetment south of Clam Creek in vicinity of fishing pier Inlet/channel migration of Brunswick River has influenced erosion along shoreline Boneyard beach lacks dune system; overwash common during storms
2	*Inlet shore to Beachview Drive and Major Horton Road* Upland elevation ranges from approximately 5 feet to 10 feet, with small scarps of eroded upland; small, young sand dunes, <5 feet, line section of northeast beach; flooding of low land probable during storms Accretion up to 160 feet occurred along small portion of section from 1857–1974 (most occurred 1924–1974); 1974–1980, approximately 60 feet of accretion occurred Recent erosion has severely affected southern portion Granite revetment along southern portion Inlet migration has directly affected northern section Overwash of existing dunes infrequent during spring tides and more frequent during storms
3	*Beachview Drive and Major Horton Road to area south of Bordon Lane* Elevation: An upland scarp exists along northern portion with upland elevations of approximately 10–20 feet; no dunes present; to the south, fronting the housing area, is a line of dunes <8 feet in height; upland elevation ranges from 10 feet to 15 feet; flooding unlikely except in major storms 1857–1974, erosion of up to 450 feet occurred; 120 feet, 1924–1974; continuous erosion, slowed or stabilized 1974–1980 Granite revetment extends throughout almost all of section; on northern end erosion has continued behind revetment Overwash of existing dunes may occur during some storms; exposed revetment often overwashed
4	*Bordon Lane to South Picnic Area* Upland relatively high, 10 feet to 15 feet, with points close to 20 feet. Flooding highly unlikely except during hurricanes. No dunes along northern two-thirds; large, extensive dunes have developed at southern end Long-term trend of minor erosion Maximum erosion of 560 feet, 1857–1974; net change less; general, long-term stability with recently accelerated erosion Concrete seawall and granite revetment line most of section. Structures exposed along upper three-quarters Overwash rare, except in a few low dunes
5	*South Picnic Area to Jekyll Point* Upland averages <10 feet. Dunes may be 15 to 20 feet high. Flooding potential low except for areas adjacent to sloughs and swales All of section has experienced continuous, rapid accretion since 1857, except for a slight erosional period, 1924–1974 No engineering structures present Stable, narrowing inlet to south Major overwash of dunes unlikely
6	*Jekyll Sound shore: Jekyll Point to Beach Creek/St. Andrews Picnic Area* Upland 5 to 10 feet, with some higher points and adjacent wetlands; shell/sand beach has no extensive dunes; flooding likely during storms Long-term (1857/68–1974) erosion; recently (1924–1974) relatively stable No major engineering structures in place; minor bulkheading at picnic area Inlet relatively stable

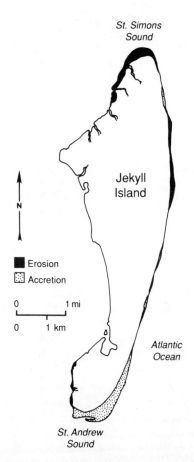

5.38 Historic net erosion and accretion: Jekyll Island, 1857/68–1974. (Adapted from Griffin and Henry, 1984.)

threatened by shoreline retreat. Development on the oceanfront is moderate risk. Construction inland is at low risk.

To the south, a series of prograding dunes marks the zone of accretion. From 1870 to 1967 the area accreted from 370 to 500 feet. The dunes are high and continuous, reducing the risk of flooding and washover. Development here is at low risk.

Elevation. Elevations range from sea level to 30 feet, among the highest in the Georgia isles. Much of Jekyll's upland is about 10 feet above MSL. Ranging in elevation between 10 and 20 feet, the island's Pleistocene core is the site of most residential areas. Beachfront developments, from houses to motels, tend to be located on land 5 to 15 feet in elevation. Some of these structures are well-protected by a healthy dune system, while others are separated from the beach by a discontinuous line of only scattered dunes. Along some stretches of shoreline, no dunes are to be found! On Jekyll's backside, the "Millionaire's Village" area and a relatively new residential area are located on land ranging from 5 to 10 feet in elevation.

Parallel dunes of Holocene sands on the south end are 10 to 30 feet high; overall average dune elevation is between 10 and 15 feet. No development has occurred in any of these areas, with the exception of a fishing pier on the north end and a small housing area and motel on the south end.

History of MHW Changes. Jekyll Island is the most stable of all the Georgia islands (as measured in net change, 1850s to 1970s). The island is currently experiencing erosion on the northwest shoreline and accretion on the southeast shoreline, resulting in an overall shift (fig. 5.38). Since 1857, erosion of 950 feet has been documented in the northwest, while accretion of 1,420 feet has been measured on the southeast shore. Erosion along the central portion ranged from 170 feet to 560 feet during that period.

St. Simons Sound Shore. The Jekyll shore of St. Simons Sound has retreated continuously at an average annual rate of 8 feet between 1857/68 and 1967. Between 1974 and 1980 the measured rate of erosion was 5 to 10 feet per year.

Oceanfront Strand. The oceanfront strand of Jekyll Island has been fairly stable. Segments of erosion and accretion alternate along the oceanfront shore.

Along the ocean beaches to the immediate south of the inlet, slight but steady accretion occurred between 1857 and 1974; much of the north-central beachfront continuously eroded. Ironically, the only beachfront residential area is located along this section of shoreline that retreated as much as 450 feet since 1857.

The central and south-central shore have histories that range from mild erosion to relative stability to minor accretion. The primary dunes along these sections and along areas to the north and south were removed in 1955 with the onset of development. Currently, recreational and lodging facilities are located on the upland along these sections.

South End. The island's south end has been building out steadily since the earliest records of 1857/68 (fig. 5.38). As is true at the north end, the rate of change has been accelerating. Between 1924 and 1957 maximum shoreline advance in this area occurred at a rate of 38 feet per year.

History of Erosion Control Efforts. Hurricane Dora in 1964 caused considerable erosion along Jekyll Island. Soon thereafter, rock rubble seawalls were installed along the central and northern shorelines to protect buildings and upland at the eventual expense of the beach (fig. 5.39). Along portions of these walls (especially in the north) erosion has continued, occurring even landward of the rubble wall. The north end picnic area was permanently closed because of the area's rapid upland erosion. In the vicinity of the convention center the rubble wall was initially covered by new dunes, but recent erosion has reexposed the rock rubble.

The 1988 *Feasibility Study*, which provided a partial basis for a subsequent Corps of Engineers document, proposes replenishing 23,500 feet of oceanfront beach. The intent is to create a 100- to 150-foot-wide MHW beach with a projected life of 8 to 10 years. Currently, there are no active efforts to pursue a replenishment project for Jekyll Island's beaches.

Inlet Migration. Neither of the inlets adjacent to Jekyll Island is a rapidly migrating one.

5.39 Jekyll Island "beach," 1985. Photo by T. Clayton.

However, the northern inlet shoreline is subject to continuous erosion.

Storm Overwash. Washover is most likely on the eroding beaches of the St. Simons Sound inlet and the north-central ocean shorelines. Erosion along the inlet's shores has resulted in the retreat of upland and marsh and the deposition of sandy washover fans in the marshes. Along some sections of the north-central beach, storm waves have overwashed the rock revetment.

Storms. Historical storm records specific to Jekyll Island are rare. However, it is known that these major 19th-century hurricanes struck: an 1804 storm that created a flood 7 feet above normal tides; a tropical cyclone in 1812; an 1824 storm that caused tremendous damage to plantations on the islands; an 1896 hurricane that left $300,000 in damage in Brunswick and $150,000 in damage on St. Simons; and the October 1898 storm that resulted in at least $1 million in damage at Brunswick and St. Simons. Notable 20th-century hurricanes have included an October 1944 storm that caused damage in Brunswick;

Selecting a Site on Georgia's Barrier Islands **105**

an October 1950 storm; a June 1957 storm during which the administration building on Jekyll was destroyed; and Hurricane Dora in September 1964. The last major storm to affect Jekyll was Hurricane David, a category 2 storm, in 1979. Wind speeds at Brunswick were clocked at 50 mph, and up to 20 feet of dune recession occurred on Jekyll.

As elsewhere, northeasters can be a problem, especially when they coincide with a high tide.

Evacuation. Jekyll Island is the last developed Georgia barrier to use a drawbridge. (The bridge's entire center span is lifted up to allow large boats and ships to pass through Jekyll Creek behind the island.) This bridge and a 6.2-mile-long two-land causeway provide the only links to the mainland.

A potential evacuation hazard is the possibility of problems with the drawbridge. If the bridge jammed or broke in the "up" position, vehicles would be stranded. Flooding of the causeway is also a potential problem, but much of the highway's elevation should delay early inundation. Despite these potential dangers, evacuation of Jekyll Island should be relatively simple and rapid. Jekyll's relatively sparse population, as well as the island's small size, mean that time is on the side of emergency management teams. The evacuation route after people arrive on the mainland could affect the speed at which Jekyll residents are able to leave the coastal area.

Services. Although Jekyll Island is in Glynn County, most of the island's services are provided by or managed by the Jekyll Island Authority. The *Jekyll Island Comprehensive Land Use Plan* gives guidelines for provision and management of services. In general, the island offers the basics: water, sewers, utilities, and some businesses.

Water use on Jekyll is currently limited by the Environmental Protection Division to less than 2.15 million gallons daily (barely enough to meet even current usage). This water, provided by the Jekyll Island Authority water system, is pumped from the Floridan (principal artesian) aquifer.

No schools or major shopping areas are located on Jekyll; however, there are three churches. Also available are basic commercial services such as food stores, restaurants, a pharmacy, gasoline stations, and a post office. In general, the insular nature of a barrier has been retained on Jekyll despite the abundance of tourist-related enterprises.

Land Use. Jekyll Island serves as a major public recreational beach for this region of the coast and supports a year-round population of about 1,500. In 1991 there were about 400 commercial and residential sites on the island.

Since the state purchased the island in 1947, management has been handled by the Jekyll Island Authority. In the early decades of state ownership, emphasis was placed on creating the image of Jekyll as a vacation resort. Only during the 1980s did natural resource management become important.

In 1983 the Institute of Community and Area Development at the University of Georgia prepared the *Jekyll Island Comprehensive Land Use Plan*, which establishes a strong, consistent strategy for the use and management of resources. The plan includes stringent guidelines for protection and enhancement of the dynamic beaches and adjacent high-hazard upland zones. In many cases the plan is stronger than existing state and county guidelines.

The following are "policy application statements" that apply to the beachfront upland.

On the north end of Jekyll: additional high investment facilities will not be allowed. Existing motels are considered to be nonconforming uses; if they are destroyed, they will not be rebuilt.

On the central shoreline: a "zone of no encroachment" is established along the beach. Within this area no new development will be allowed; parking areas have been removed.

On the south end beachfront area: no additional development of residences or businesses is to occur. Structures in the existing residential area should not be rebuilt if destroyed; vacant lots should not be leased.

Current uses of Jekyll will continue; the Authority cannot sell or lease more than 35 percent (about 400 acres) of the island above the mean high tide line. Land will be leased until the year 2049 when the Authority is to be dissolved. (At that time the state will continue to manage the island; however, no detailed plans seem to have been formulated.)

As of early 1991 the Land Use Plan was being revised and rewritten by the Institute for Community Development. According to officials within the Authority, the new plan was to include even more stringent policies regarding natural resource and land-use management.

Little Cumberland Island

Little Cumberland (figs. 5.40 and 5.41; table 5.12) is a typical Holocene island composed of a series of fantail-shaped beach ridges. It is 2.4 miles long with an area of 1,410 acres.

Elevation. The island's elevation ranges from sea level to about 40 feet. The north and northeast portions of the shoreline bordering the Cumberland River and St. Andrew Sound have forested dunes perpendicular to the shore. Although dunes oriented in this manner are not as protective as dunes parallel to the shoreline, the large quantity of sand contained in the system serves as a significant buffer against storm wave attack.

The island's east-facing section also is backed by high ground. The upland landward of Christmas Creek is bounded by tidal marsh on both the ocean- and soundsides. The elevation here is somewhat lower than on the main island, and some flooding may occur during severe storms.

History of MHW Changes. Forest vegetation up to the Cumberland River/St. Andrew Sound shoreline edge indicates that erosion predominates over accretion in this section. Records also show consistent shoreline recession along the island's west/northwestern shore (fig. 5.41). The Cumberland River shore has been steadily erosive since 1857/68. On the north end of the island, the northwest corner shows net erosion from 1857/68 to 1974, while the northeast corner experienced net accretion.

Historically, the east-facing shore has shown great instability. Physiographic and vegetative patterns, historical maps, charts, and photographs all indicate the area is highly dynamic, with both erosion and accretion occurring along this section of beach. Because this section is located near the northward-migrating Christmas Creek and St. Andrew Sound, rapid shoreline changes should be expected.

The section landward of Christmas Creek and the Christmas Creek Inlet is strongly influenced by the inlet. This area is bounded by tidal marsh on both the ocean- and soundsides. Meandering of Christmas Creek has truncated this section's seaward edge.

Inlet Migration. The St. Andrew Sound inlet is relatively stable, although its shores are subject to significant change. The Christmas Creek inlet is an extremely active tidal inlet; between 1974 and 1982 it migrated northward about 1,700 feet.

Evacuation. Little Cumberland Island is accessible by boat only; therefore, evacuation must begin early in the event of an approaching storm.

Land Use. The island is owned by the Little Cumberland Island Association and is incorporated into the Coastal Barrier Resources System. Three-quarters of the island is designated wilderness, while the remaining one-quarter is open to limited deed-restrictive development. The island currently has mostly low-density development with large individual lots. Homes are generally located some distance from the beach and are in little danger from erosion.

Cumberland Island

The National Park Service manages most of Cumberland Island (fig. 5.41) as the Cumberland Island National Seashore. However, some scattered holdings are controlled by private interests.

The book *Encounters with the Archdruid* by John McPhee contains an interesting account of the encounter between developer Charles Fraser and conservationist David Brower in Greyfield Inn on Cumberland Island. This fascinating story provides historical background on how the Cumberland Island National Seashore was established. For another provocative discussion of the conflicting demands of conservation, recreation, and development, see Anne Simons's book, *The Thin Edge: Coast and Man in Crisis* (appendix D). She uses Cumberland Island, part of the country's first national seashore, as a case study.

The island's general shoreline history indicates that the northern quarter has experienced accretion during the past 50 years. The central portion is stable, exhibiting little shoreline

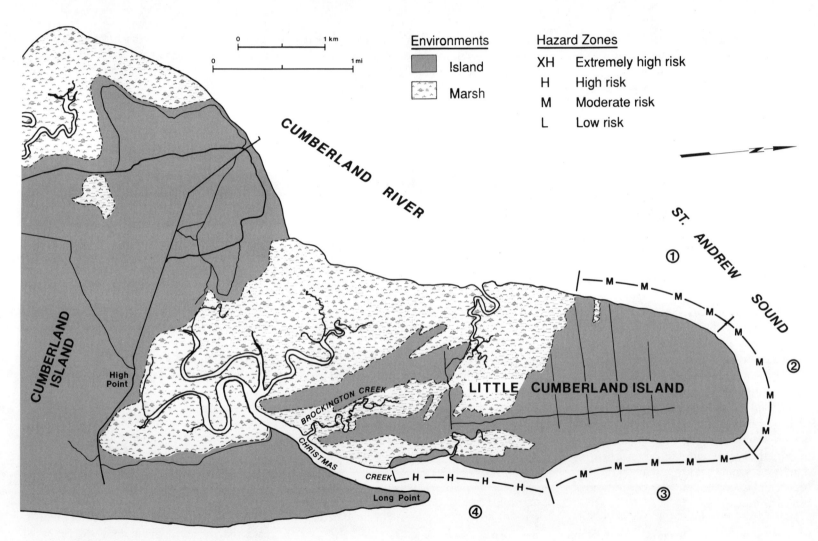

5.40 Site analysis map: Little Cumberland Island. Companion map to table 5.12.

108   Living with the Georgia Shore

Table 5.12 Little Cumberland Island Shoreline Hazard-Risk Evaluation

Shoreline section	Comments
1	*Cumberland River Shoreline* Upland elevation ranges from <5 feet adjacent to marshes and shoreline to 10–15 feet along crests of parallel dune-ridges. Storm flooding likely along marsh edges Net erosion, 1868–1974, along entire section; accretional trend began in 1957 with up to 85 feet of landward growth through 1982 Cumberland River channel and inlet migration affect sediment movement
2	*St. Andrews Sound inlet shoreline* Upland adjacent to inlet unusually high (25–50 feet) with approximately 10-foot-high land east of and behind this ridge Flooding of lower areas likely during major hurricanes Subject to cycles of erosion and accretion. 1868–1974, west side experienced net erosion, and northeast/ocean portion experienced net accretion. 1974–1982, accretionary trend reversed
3	*Inlet shore to central area of ocean shore (north of spit)* 10- to 15-foot parallel ridges line upland with scattered points >15 feet. Flooding likely only in areas <5 feet high near marshes and swales during major storms Continuous erosion since 1868 Subject to influence of both St. Andrews Sound inlet and Christmas Creek inlet Dune overwash improbable except during major storms
4	*Central ocean beach to Brockington Creek and Christmas Creek* Upland elevation <5 feet along marsh; scattered ridges of approximately 10 feet. Flooding may occur on high storm tides in upland adjacent to marsh. Accretion along this shoreline with minor exceptions associated with occasional breaching of Long Point spit and relocation of Christmas Creek Christmas Creek inlet actively migrates Overwash of dunes probable during storms and some spring tides

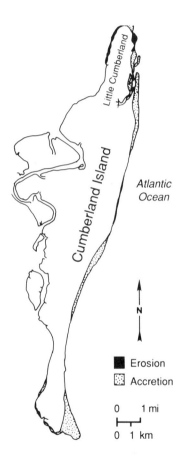

5.41 Historic net erosion and accretion: Little Cumberland Island and Cumberland Island, 1857/68–1974. (Adapted from Griffin and Henry, 1984.)

change over the past half-century. Considerable accretion has occurred on the southern 7 miles, in part because of construction of the north jetty at the St. Marys River.

Greyfield and Stafford, two privately owned tracts, include shorefront lands; however, these sites are located where erosion is not a threat. High ground near the island's northeastern end is susceptible to erosion through inlet formation or migration of the existing inlet, Christmas Creek.

In many respects, Cumberland may be the most precious jewel in the necklace of islands that adorns Georgia's coast. As the state's largest undeveloped island, Cumberland exhibits the greatest variety of natural habitats and systems. Traces of each stage in the history of Georgia's islands, from Native Americans to protection of public land, are to be found. And, perhaps most important, the island is undeniably imbued with the magic unique to Georgia's golden isles.

The Shore Challenge

Your choice of a particular island and site, of course, will consider obvious physical factors such as shoreline erosion and the presence or absence of shade trees. Shorefront shoppers, however, sometimes forget to take political factors into account. How do the state and community manage the shoreline? Are structures adjacent to your property well-built, conforming to code? What future development is planned? Will single-family homes give way to condos or high-rises? How are long-range growth plans to be implemented? Are planning documents presented as suggestions and springboards for discussion, or are there regulatory means in place to implement them? Is the environment you've come to enjoy (beaches, dunes, marshes, maritime forests, the fisheries) protected from the detrimental effects of coastal development (hard stabilization, habitat destruction, stormwater runoff, etc.)?

Management is generally more important in the coastal zone than at most inland sites. Chapter 6 outlines some of the legal framework within which coastal dwellers and managers must work. Bear in mind that the red tape of required permits is more than a nuisance; it reflects efforts to protect the natural and recreational resources that all of us enjoy. Your participation in local planning can help to ensure the perpetuation of those charms.

6 Coastal Land-use Planning and Regulation

During the 1970s and 1980s land-use planning and regulation in the State of Georgia's coastal zone increased substantially. A major milestone in this change in the way lands and waters may be used was passage of the state Coastal Marshlands Protection Act of 1970, regulating filling, dredging, and other alterations to marshlands within the estuarine area. To reduce erosion losses along the state's barrier islands, another regulatory program was established under the Shore Assistance Act of 1979.

Unlike most coastal states, Georgia is not participating in the federal coastal zone management program. The state's Coastal Management Act of 1978, intended to enable the state to seek financial assistance under the federal program, was repealed in 1984. Important federal programs having an impact in Georgia include the National Flood Insurance Program and the U.S. Army Corps of Engineers' dredging and filling permit system.

A discussion of some of the relevant state and federal land-use and regulatory programs applicable to the Georgia coastal zone follows. The explanations are introductory and general in nature; further information may be obtained from the agencies administering the programs (appendix C). Appropriate county and city offices should be contacted for local planning, zoning, and building ordinances and regulations.

Coastal Marshlands Protection Act of 1970

To achieve a balance between "protection of the environment on the one hand and industrial and commercial development on the other" (1970 Ga. Laws 940), the Georgia General Assembly enacted the Coastal Marshlands Protection Act in 1970 (Ga. Code Ann. ch. 43-2401).

This law requires that permits be obtained before any alterations are made in the "marshlands" within the state's "estuarine area." The "estuarine area" is defined as "all tidally influenced waters, marshes, and marshlands lying within a tide-elevation range from 5.6 feet above mean tide level and below" [Ga. Code Ann. § 43-2402(4)]. The term "marshlands" as used in the act means "any marshland or salt marsh . . . , whether or not the tidewaters reach the littoral areas through natural or artificial water courses," and includes "those areas upon which grow salt marsh grass . . . , black needlerush . . . , and high-tide bush . . ." [Ga. Code Ann. § 43-2402(2)]. By regulation, additional types of vegetation are included within the definition.

The act created the Coastal Marshlands Protection Committee, which is empowered to issue orders and to grant, deny, revoke, and amend permits. The committee consists of the Commissioner of Natural Resources and two persons selected by the Board of Natural Resources. The state Department of Natural Resources administers and enforces the act and the rules and regulations promulgated under it.

When Do I Need a Permit?

If you want to "remove, fill, dredge, drain, or otherwise alter any marshlands . . . within the estuarine area," you must apply for and obtain a permit from the Coastal Marshlands Protection Committee (Ga. Code Ann. § 43-2405). However, the law does not require permits for certain activities, including "the building of private docks on pilings, the walkways of which are above the marsh grass not obstructing tidal flow, by the owners of residences located on high land adjoining such docks" [Ga. Code Ann. § 43-2412 (6)].

What Must Be in the Application?

You may obtain the application to alter coastal marshlands from the state Department of Natural Resources (see appendix C for address). You must include a copy of your deed or, if you're not the owner, a copy of the owner's deed together with the owner's written permission to carry out the project on his land. You must accompany your application with a certified check or money order for $25 for each acre of land or portion of an acre; it is payable to the Georgia Department of Natural Resources.

Your application must contain a complete

description of the proposed activity, including drawings, sketches, or plans, and the location, purpose, and intended use of the proposed activity; scheduling of the activity; and the names and addresses of adjoining property owners and the location and dimension of adjacent structures. It must set forth any approvals or denials already made by federal, state, or local agencies for the work.

If your proposed activity involves dredging, your application must describe the type, composition, and quantity of the material to be dredged, the method of dredging, and the site of and plans for disposal of the dredged material.

You may be required to furnish additional information to assist in evaluating your application. Applicants are also encouraged to present, in writing, reasons why the permit should be granted.

How Is the Application Processed?

Within 15 days of receipt of a completed marshlands application, the state notifies all adjoining landowners in writing about the proposed use or activity. If the applicant has stated that the names or addresses of the adjoining owners could not be ascertained, the state, within 30 days of receipt of the completed application, must publish a legal notice of the proposed activity and a brief description of the land to be affected.

Applications are heard by the three-member Coastal Marshlands Protection Committee, which usually meets on the second Tuesday of every month in one of the state's coastal counties or in Atlanta. Written notification of the date, time, and place of committee meetings must be given to applicants for permits and to people filing statements relating to the applications to be heard. Special meetings may be called by any committee member upon seven days' advance notice to the other committee members, applicants whose applications will be heard, and the public.

Guidelines for Permit Evaluations

In evaluating applications for permits, the committee follows guidelines set forth in its rules and regulations. These rules define what is deemed to be in the public interest, and furnish general and specific guidelines for alteration of marshlands.

As the applicant, you must demonstrate to the committee that the proposed alteration is not contrary to the public interest. A permit will not be approved if the committee finds that an application is contrary to the public interest.

Under the guidelines, these three questions must be considered: (1) Will the proposal result in any unreasonably harmful obstruction to or alteration of the natural flow of navigational water within the area? (2) Will the proposal cause unreasonably harmful or increased erosion or shoaling of channels or create stagnant areas of water? (3) Will the proposal unreasonably interfere with the conservation of fish, shrimp, oysters, crabs, and clams or any marine life or wildlife or other natural resources?

Ordinarily, a permit will not be granted if the proposed alteration "is not water-related or dependent on waterfront access, or can be satisfied by the use of existing public facilities." You, as the applicant, must show that a feasible alternative site does not exist, and you are encouraged to show an overall public interest for the proposal.

The general guidelines provide that "in most cases, only those marine-oriented activities and structures which must have a shoreline or marshlands location in order to function will be considered" favorably. In any event, the guidelines require that "the amount of marshlands to be altered should be minimum in size." Permits are more likely to be granted for the following activities and structures: (1) marinas and public docks where the demand for the facilities justifies the alteration of marshlands; (2) channel dredging essential for navigation and public health purposes; (3) public recreation; (4) erosion control structures; (5) enhancement of public access to the water by the installation of fisherman's catwalks, boat-launching ramps, or other structures; and (6) roads placed on pilings rather than built as solid-fill causeways.

The guidelines state that the following activities and structures are generally considered to be contrary to the public interest when located in marshlands: (1) filling for residential, commercial, and industrial uses; (2) filling for private parking lots and private roadways; (3) boat storage sites; (4) dump sites and depositing of waste materials or spoil; (5) dredging of canals or ditches solely for the purpose of draining marshlands; (6) mining; (7) lagoons or impoundments for waste treatment, cooling, or aquaculture that would occupy or damage significant ecologically productive marshlands; and (9) structures that would unreasonably obstruct the adjoining landowners' views.

There are special guidelines for each of the activities and structures that are more likely to be approved. For example, in considering whether erosion control structures should be approved, the rules and regulations provide these guiding principles: construction of bulkheads, groins, jetties, and other shore and erosion control structures will be approved only in areas with unstable shorelines and serious erosion problems; erosion control structures will be discouraged where marshlands are adequately serving as an erosion buffer or where public access would be adversely affected; and riprap and/or designs using natural vegetation will be encouraged instead of wood, concrete, or metal bulkheads.

Decisions on Permit Applications

A majority vote of the Coastal Marshlands Protection Committee is required either to approve or deny an application to alter marshlands. The committee must inform the applicant in writing as soon as possible after the meeting of any action it takes. In any event, the committee must act within 90 days of the filing of a completed application.

The granting of a permit may be conditioned upon the applicant's amendment of the proposal to take whatever measures are necessary to protect the public interest. When a committee member determines that a conditional permit should be issued, that member must notify the other members in writing, and the committee has an additional 15 days to act on the application.

Appeal and Enforcement Procedures

Under the committee's rules and regulations, "any person who is aggrieved or adversely affected by order or action of" the Department of Natural Resources has the right to a hearing. The request for such a hearing must be filed within 30 days of the order or notice of such action. After exhausting all administrative remedies within the DNR, the applicant who is dissatisfied with a final order may seek review in the courts.

To enforce the act or any orders, the DNR is authorized to use one or any combination of these methods: (1) It may issue a cease-and-desist order or other appropriate order whenever someone is altering the marshlands within a permit, is doing so in violation of the terms and conditions of a permit, or is violating the provisions of the act in any other manner. (2) Whenever the DNR, after a hearing, determines that anyone has failed or refused to comply with any provision of the act or any order of the DNR, it may impose a civil penalty up to $1,000 and an additional civil penalty of up to $500 for each day during which the violation continues. (3) The DNR may have the superior court enter a judgment upon filing either a final order of the DNR that was not appealed from, or a final DNR order affirmed upon appeal. (4) The DNR may seek injunctive or other relief in the superior court.

Term and Transfer of Permit

Although a permit to alter marshlands becomes final immediately upon being issued, no construction or alteration may begin until 45 days after the committee's approval or, if an appeal is filed on time, until all administrative and judicial proceedings have been completed. If the project is not completed within two years of the permit's issuance, the permit will become void. However, upon a showing of good cause, the time limit may be extended. The committee, after notifying the permit holder of its intention, may revoke a permit for noncompliance or violation of the terms of the permit.

If the person receiving a permit sells, leases, rents, or otherwise conveys the property, the permit remains effective in the name of the new owner, lessee, tenant, or other assignee, provided there is no change of the use of the land as set forth in the original application. The original applicant must notify the committee in writing within 30 days of the conveyance.

Shore Assistance Act of 1979

Sand dunes, beaches, and offshore bars and shoals—which make up a "sand-sharing system"—are within the jurisdiction of the Shore Assistance Act of 1979 (Ga. Code Ann. ch. 43-30). Among legislative findings enumerated in this law are the following: "the sand-sharing system . . . acts as a buffer to protect . . . property and natural resources from the damaging effects of floods, winds, tides, and erosion"; "the coastal sand dunes . . . are easily disturbed by action harming their vegetation or inhibiting their natural development"; "[r]emoval of sand from [offshore sandbars] and shoals can interrupt natural sand flows and have unintended, undesirable, and irreparable effects"; and "this natural resource system is costly, if not impossible, to reconstruct or rehabilitate once adversely affected by man-related activities and is important to conserve for the present and future use and enjoyment of all citizens and visitors to this state . . . and is an integral part of Georgia's barrier islands, providing great protection to the state's marshlands and estuaries" (Ga. Code Ann. § 43-3002).

Under this law, permits are required for certain structures, "shoreline engineering activities," and land alterations. The act is applicable in the "dynamic dune fields" on the barrier islands and in the adjoining "submerged shoreline lands" (Ga. Code Ann. § 43-3004). The term "dynamic dune field" is defined as "the dynamic ocean-facing area of beach and sand dunes, varying in height and width, the ocean boundary of which extends to the ordinary high-water mark and the landward boundary of which is the first occurrence either of live native trees 20 feet in height or greater, of coastal marshlands. . . , or of an existing structure" [Ga. Code Ann. § 43-3003(8)]. The term "submerged shoreline lands" means "the intertidal and submerged lands from the ordinary high-water mark seaward to the limit of the state's jurisdiction in the Atlantic Ocean" [Ga. Code Ann. § 43-3003(20)].

The act is administered by the Shore Assistance Committee, the state Department of Natural Resources, and the counties and municipalities in which the dunes are located. As mentioned, this three-member committee consists of the Commissioner of Natural Resources and two persons from the coastal counties. The latter two are selected by the Board of Natural Resources. The committee has exclusive permitting authority for shoreline engineering activities and for activities proposed to occur in whole or in part on submerged shoreline lands or on other state-owned lands. Local governments that have enacted the requisite ordinances and have been certified by the board are the permit-issuing authorities for other types of permits required under this act.

When Do I Need a Permit?

If you want to "construct or erect any structure or construct, erect, conduct, or engage in any shoreline engineering activity or engage in any land alteration which alters the natural topography or vegetation of any area within the jurisdiction of" the Shore Assistance Act, you must obtain a permit (Ga. Code Ann. § 43-3005). "Shoreline engineering activity" is broadly defined to include such activities as grading, excavating, artificial dune construction, beach nourishment, and the construction and maintenance of groins, seawalls, jetties, pipelines, and piers [Ga. Code Ann. § 43-3003(17)].

What Must Be in the Application?

Applications for permits must be on forms prescribed by the state Shore Assistance Committee or by the county or municipality whose shore assistance program has been approved by the state Board of Natural Resources, and you must pay the application fee prescribed by the permit-issuing authority. You must include a copy of your deed or, if you're not the owner, a copy of his deed along with written permission from the owner to carry out the project on his land.

You must briefly describe the proposed project and include a plat depicting the boundaries of the site and site plans showing proposed streets, utilities, buildings, and any other physical structures. You must set forth the names and addresses of all adjoining landowners. Your application must include an architect's or engineer's certification that any proposed structures are designed to meet the Board of Natural Resources' standards for hurricane-resistant buildings.

How Is the Application Processed?

Within 10 days of receipt of a completed application, the permit-issuing authority notifies all adjoining landowners about the proposed project. In evaluating applications, certain assessment tools and techniques are used. Among these are historical photographs and topographic data, on-site inspections, and climatological and meteorological records. Action must be taken on a permit application within 60 days after it is completed unless the applicant waives this requirement in writing.

Guidelines for Permit Evaluations

Permits for structures or land alteration are issued only when (1) the proposed project will occupy the landward area of the subject parcel; (2) a "reasonable percentage" of the parcel will be kept in its naturally vegetated condition; (3) the activities associated with the proposed project are kept to a minimum, are temporary in nature, and, upon completion of the project, will restore the natural topography and vegetation to at least its former stability; and (4) the proposed project will maintain the normal functions of the sand-sharing mechanisms in minimizing storm-wave damage and erosion.

Except for shoreline engineering activities, boardwalks, or crosswalks, permits will not be issued for a structure or land alteration on beaches, eroding sand dune areas, and areas without stable sand dunes. Permits will be issued for shoreline engineering activities or for land alterations or structures on submerged shoreline lands only under certain conditions.

Guidelines for permit evaluations are contained in the rules and regulations adopted under the Shore Assistance Act. There are general guidelines for permits involving use of lands located in dynamic dune fields, beaches, eroding sand dune areas, and areas without stable dunes, as well as for shoreline engineering activities and activities on submerged shoreline lands.

Specific guidelines are set forth for fishing piers and other recreational structures, which must be built so that water flow is not restricted. In considering erosion control structures, beach nourishment techniques are to be preferred, and permits for vertical seawalls, bulkheads, and riprap will be granted only when the applicant has demonstrated that no reasonable or viable alternative exists.

Shore Assistance Committee Meetings

The committee meets at least once a month if there are completed applications for permits on file. Its meetings are held either in one of the coastal counties or in Atlanta. Applicants are notified of the time and place of meetings, and they and other interested persons may present their views at such meetings. A majority vote of the committee is required to grant or deny applications.

Appeal and Enforcement Procedures

Anyone who is aggrieved or adversely affected by any order or action of the Shore Assistance Committee has the right to a hearing before an administrative law judge appointed by the Board of Natural Resources. The decision of the administrative law judge may be reviewed in court. Actions of local governmental units administering the law are also subject to administrative and judicial review.

If the Department of Natural Resources determines that anyone is violating this law, any of the applicable rules and regulations, or the terms and conditions of any permit, and the violation is in an area where the committee is the permit-issuing authority, the DNR may use any or all of three enforcement methods. (1) It may issue an administrative cease-and-desist order and require that corrective action be taken, such as returning sand dunes, beaches, and submerged shorelines to the condition they were in before the violation. (2) Under certain conditions, it may issue an emergency order requiring such action as it deems necessary to meet the emergency. (3) It may seek injunctive relief in court.

The law provides that activities in violation of this act or any ordinance or regulation adopted pursuant to it are deemed public nuisances that may be enjoined or abated by a superior court action.

National Flood Insurance Program (NFIP)

The National Flood Insurance Act of 1968 (P.L. 90-448), as amended by the Flood Disaster Protection Act of 1973 (P.L. 93-234), is intended to encourage prudent land-use planning and to minimize property damage in flood-prone areas, including the coastal zone. To participate in the National Flood Insurance Program, local communities must adopt ordinances to reduce future flood risks. The NFIP provides an opportunity for property owners to purchase flood insurance that generally is not available from private insurance companies.

The initiative for participating in the NFIP rests with the community, which must get in touch with the Federal Emergency Management Agency (FEMA). Any community may join the National Flood Insurance Program provided that it adopts and enforces NFIP minimum flood plain management standards for regulating all proposed construction and other development within the flood zone and ensures that construction materials and techniques are used to minimize potential flood damage. The federal government makes a limited amount of flood insurance coverage available, charging subsidized premium rates for all existing structures and/or their contents, regardless of the flood risk. All new construction and substantially improved existing structures are charged actuarial insurance rates.

FEMA provides a detailed Flood Insurance Rate Map (FIRM) indicating flood elevations and flood-hazard zones, including velocity zones (V-zones) for coastal areas where wave action is an additional hazard during flooding. The FIRM identifies Base Flood Elevations (BFES), establishes special flood-hazard zones, and provides a basis for floodplain management and establishing insurance rates.

To participate in the NFIP, the community must adopt and enforce management ordinances that meet at least the minimum requirements for flood-hazard reduction as set by FEMA. All new structures will be rated on an actual risk (actuarial) basis, which may mean higher insurance rates in coastal high-hazard areas, but generally results in a savings for development within numbered A-zones (areas flooded in a 100-year coastal flood, but less subject to turbulent wave action).

FEMA maps commonly use the "100-year flood" as the base flood elevation to establish regulatory requirements. People unfamiliar with hydrologic data sometimes mistakenly take the "100-year flood" to mean a flood that occurs once every 100 years. In fact, a flood of this magnitude could occur in successive years, or twice in one year, and so on. If we think of a 100-year flood as a level of flooding having a 1 percent statistical probability of occurring in any given year, then during the life of a house within this zone that has a 30-year mortgage, there is a 25 percent probability that the property will be flooded. The chances of losing your property becomes 1 in 4, rather than 1 in 100. Having flood insurance makes good sense.

In V-zones new structures are subjected to velocity water and wave action, a risk factor over and above other flood hazard areas. Elevation requirements are adjusted, usually 3 to 6 feet above still-water flood levels, for structures in V-zones to minimize wave damage, and the insurance rates also are higher. When your insurance agent submits an application for a building in a special flood hazard area, a certificate that verifies the elevation of the building's first floor must accompany the application.

The insurance rate structure provides incentives of lower rates if buildings are elevated above the minimum federal requirements. Flood insurance coverage is provided for structural damage as well as contents. Most coastal communities are participating in the NFIP. To determine if your community is in the NFIP and for additional information on the insurance, get in touch with your local property agent or call the NFIP's servicing contractor (phone: (404) 853-4400), NFIP Region IV Office at 1371 Peachtree Street, N.E., Atlanta, Georgia (appendix C). For more information, request a copy of "Answers to Questions About the National Flood Insurance Program" from FEMA.

During the spring of 1991 the U.S. House of Representatives passed legislation that would make significant changes in the National Flood Insurance Program. At the time of this writing, the legislation still needs to be passed by the Senate and then signed into law by the president. This legislation would establish a new program to reduce coastal erosion hazards along U.S. tidal waters and Great Lakes shorelines. The proposed program, which will be part of the NFIP, would include identification of erosion-prone communities; erosion setbacks; required adoption of land-use restrictions in erosion-prone communities; and erosion mitigation assistance to owners of structures determined to be eligible for this assistance. The NFIP legislation also proposes changes to the purchase of repetitive flooded property program and increases in the amount of insurance coverage being made available. This proposed legislation indicates a growing recognition in Congress that the nationwide problem of coastal erosion needs to be addressed. If this legislation is not passed, we can anticipate that similar legislation will be introduced in the future.

Before buying or building a structure on the coast, an individual should ask certain basic questions:
–Is the community participating in the National Flood Insurance Program?
–Is my building site above the 100-year flood level? Is the site located in a V-zone? (V-zones are high-hazard areas and pose serious problems.)
–What are the minimum elevation and structural requirements for my building?
–What are the limits of coverage?

Make sure your community is enforcing the ordinance requiring minimum construction standards and elevations. Most lending institutions and local governmental planning, zoning, and building departments will be aware of the flood insurance regulations and can provide assistance. It would be wise to confirm such information with appropriate insurance representatives. Any authorized insurance agent can write and submit a National Flood Insurance policy application. All insurance companies charge the same rates for National Flood Insurance policies.

Coastal Barrier Resources Act of 1982 and Coastal Barrier Improvement Act of 1990

Recognizing the serious hazards, costs, and problems with federally subsidized development of barrier islands, the U.S. Congress passed the Coastal Barrier Resources Act (P.L. 97-3348) in 1982 and reauthorized the law by the Coastal Barrier Improvement Act of 1990 (P.L. 101-591). The purpose of these federal laws is to minimize the loss of human life and property, wasteful expenditure of federal taxes, and damage to fish, wildlife, and other natural resources from incompatible development along the Atlantic and Gulf Coasts.

Any structure built on the designated barrier islands after October 1, 1983, is not eligible for federal flood insurance. Although the law does not prohibit private development on the designated barrier islands, it passes the risks and costs of development from taxpayers to owners. With certain exceptions, the act prohibits the expenditure of federal funds, including loans and grants, for the construction of new roads, bridges, causeways, and other infrastructures that encourage barrier island development. Expenditure of funds on some projects, such as existing channel improvements and public roads that are essential links in a larger system, are permissible under the act.

Georgia currently has 33,100 acres of barrier lands within the Coastal Barrier Resources Act (CBRA) system. These areas are located on

Little Tybee, Wassaw, Little St. Simons, Sea, Little Cumberland, and Cumberland Islands. The Coastal Barrier Improvement Act of 1990 expanded the Coastal Barrier Resources system for the United States. No new units were added in Georgia; however, a 3.2-mile (5.2-kilometer) shoreline length increase was recommended and would add over 31,000 acres to the system. These new areas are tracts of wetland acreage adjacent to barriers already included in the CBRA system. The proposed additions would bring Georgia's CBRA total to more than 64,000 acres of land.

Other Federal Programs

Several other federal programs are applicable to Georgia's coastal zone. Coastal residents, property owners, officials, and developers should be aware of them.

The U.S. Army Corps of Engineers is the federal agency responsible for regulating dredging and filling along the Georgia coast. It is charged with administering the Rivers and Harbors Act of 1889 (33 USC 403) and the federal Water Pollution Control Act of 1972 (P.L. 92-500), as amended. If your project is within the Corps' jurisdiction, you must obtain one of these authorizations: (1) a nationwide permit, which is a general permit issued by the Chief of Engineers in 1982; (2) a regional permit, which is a general permit issued by the Savannah District of the Corps authorizing certain minor activities in specific geographic areas within Georgia; or (3) an individual permit, which must be applied for if your project exceeds the scope of the nationwide and regional permits.

A joint application form is used to apply for both the Corps permit and the state coastal marshlands permit. The Corps welcomes preapplication inquiries and particularly encourages such contact during preliminary planning for commercial, industrial, public service, and recreational projects. You may telephone the Regulatory Branch of the Savannah District office toll-free: 1 (800) 241-3715 (in Georgia only); the regular number is (912) 944-5347.

Other federal laws with a major impact on Georgia's coastal zone are administered by the Environmental Protection Agency, the Fish and Wildlife Service, and the National Marine Fisheries Service.

Georgia is the only Atlantic Coast state not participating in the federal coastal zone management program. The program, developed under the federal Coastal Zone Management Act of 1972 (P.L. 92-583), as amended, is administered by the Office of Coastal Resources Management, National Oceanic and Atmospheric Administration, Department of Commerce.

In addition to the laws noted above, other federal regulations may be important locally. Coastal residents should check with the state Department of Natural Resources or local governmental planning, zoning, and building departments.

Summary

While Georgia, unlike other Atlantic Coast states, is not participating in the federal coastal zone management program, the state has initiated widespread regulation of the lands and waters near and along the coast during the past two decades. The permit programs created under the Coastal Marshlands Protection Act of 1970 and the Shore Assistance Act of 1979 are the state's two major regulatory programs in the coastal zone. The National Flood Insurance Program and the U.S. Army Corps of Engineers' dredging and filling permit system are among important federal programs having an impact on coastal land use.

With continuing population growth in Georgia, coupled with concerns for the environment, additional land-use planning and regulation may be expected along the state's coast in the future.

Recent Additions

In 1991, the Department of Natural Resources rewrote the Coastal Marshlands Protection Act of 1970 and the Shore Assistance Act of 1979. The state legislature will consider these changes in early 1992. The proposed modifications affect definitions of terms, permitting requirements, and many other important aspects of state legislation. Be sure to obtain the most recent versions of these acts.

The recently passed Growth Management Strategies Act for Georgia includes legislation that will guide future development and resource management throughout the state. In coastal Georgia the CGRDC will probably be responsible for the development and implementation of the coastwide plan required by this legislation. This act offers an excellent opportunity for instituting coordinated management of resources in Georgia's coastal zone.

In early 1992, the state expressed interest in joining the federal Coastal Zone Management program. For current information, contact the Coastal Resources Division of the Department of Natural Resources.

Another issue of current concern is the recent redefinition of "wetlands" in the *Federal Manual for Identifying and Delineating Jurisdictional Wetlands*; these changes have in fact created a storm of controversy. Under the revised manual definitions, many areas not previously subject to wetlands regulations, such as pine barrens, would be subject to stricter wetlands rules. However, in the summer of 1991 federal agencies seemed to be turning to a definition of "wetlands" much narrower than even the original *Manual*. For current information, contact the Savannah District office of the U.S. Army Corps of Engineers.

7 Building or Buying a House Near the Shore

Life's important decisions are based on an evaluation of the facts. Few of us buy goods, choose a career, take legal, financial, or medical actions without first evaluating the facts and seeking advice. In the case of coastal property, two general areas should be considered: site safety and the integrity of the existing or proposed building relative to the forces to which it will be subjected. Even the lowest-risk site may not provide security against storm or flood for a dwelling that is not properly constructed.

Coastal Realty Versus Coastal Reality

Coastal property is not the same as inland property. Do not approach it as if you were buying a lot in a developed part of the mainland or a subdivided farm field in the coastal plain. The previous chapters illustrate that Georgia's shores are composed of variable environments and are subjected to nature's most powerful and persistent forces. The reality of the coast is its dynamic character. Property lines are an artificial grid superimposed on this dynamism. If you choose to place yourself or others in this zone, prudence is in order.

A quick glance at the architecture of many coastal buildings provides convincing evidence that the reality of coastal processes was rarely considered in their construction. Not too many years back, old-timers wisely lived back from the shore and behind the protection of sand dunes if they were present. More recently, newcomers to the shore have built precariously close to the bluff edge, or with footings on spongy marsh subsoil, or on top of dunes as if they hope to get a better view of future storms.

Many structures along the coastline were built before the adoption of a standardized building code or Federal Insurance Administration (FIA) guidelines. Town building officials enforced, to varying degrees, whatever codes were used; wide disparity in the safety of these structures is the result. New construction and major renovation projects must adhere to FIA guidelines, but many existing dwellings were built before these requirements became effective.

This chapter provides an introduction to building near the shore, or what to look for in an existing house or building. Emphasis is on the structure itself, whether a cottage, condominium, or mobile home. Keep in mind, however, that this is a general summary—application varies with region, type and intensity of hazards, type of shore material, and even social and political factors (for example, density of development, quality of adjacent structures, and enforcement of building codes and zoning ordinances).

For those who want to learn more about construction near the shore, we recommend the book *Coastal Design: A Guide for Builders, Planners, and Home Owners*, which gives more detail on coastal construction and may be used to supplement this volume. In addition, the Federal Emergency Management Agency's *Coastal Construction Manual* is informative on coastal construction and contains additional reference material. Those considering building in areas threatened by both flooding and wave energy, especially on barrier islands or low areas where storm surge is likely, should obtain *Elevated Residential Structures* from FEMA. These and other useful references are listed in appendix D.

The Structure: Concept of Balanced Risk

A certain chance of failure for any structure exists within the constraints of economy and environment. The objective of building design is to create a structure that is both economically feasible and functionally reliable. A house must be affordable and have a reasonable life expectancy, free of being damaged, destroyed, or wearing out. To obtain such a house, a balance must be achieved among financial, structural, environmental, and other special conditions. Most of these conditions are heightened on the coast—property values are higher, there is a greater desire for aesthetics, the environment is more sensitive, the likelihood of storms is increased, and there are more hazards with which to deal.

The individual who builds or buys a house in an exposed area should understand the risks involved and the chance of harm to home and family. The risks should then be weighed against the benefits to be derived from living at this location. Similarly, the developer who is putting up a motel should weigh the possibility of destruction and death during a hurricane versus the money and other advantages to be gained from such a building. Then and only then should construction proceed. For both the homeowner and the developer, proper construction and location reduce the risks involved.

The concept of balanced risk should take into account the following fundamental considerations:

(1) A coastal structure, exposed to high winds, waves, or flooding, must be stronger than a structure built inland.

(2) A structure with high occupancy, such as an apartment building, should be safer than one with low occupancy such as a single-family dwelling.

(3) A building that houses elderly or sick people should be safer than a building housing able-bodied people.

(4) Because construction must be economically feasible, most homeowners cannot achieve total safety on the coast.

(5) A building with a planned long life, such as a year-round residence, should be stronger than a building with a planned short life, such as a mobile home or a summer cottage. Structures can be designed and built to resist all but the largest storms and remain affordable.

Structural engineering is the designing and constructing of buildings to withstand the forces of nature. It is based on a knowledge of the forces to which the structures will be subjected, and an understanding of the strength of building materials. The effectiveness of structural engineering design was reflected in the aftermath of Typhoon Tracy, which struck Darwin, Australia, in 1974: 70 percent of the housing not based on structural engineering principles was destroyed and 20 percent was seriously damaged. Only 10 percent of the nonstructurally engineered housing weathered the storm. In contrast, more than 70 percent of the structurally engineered, large commercial, government, and industrial buildings came through with little or no damage, and fewer than 5 percent of such structures were destroyed. Because housing accounts for more than half of the capital cost of the buildings in Queensland, the state government established a building code that requires standardized structural engineering for houses in storm-prone areas. The improvement has been achieved with little increase in construction and design costs.

Coastal Forces: Design Requirements

Wind

Hurricanes, with their associated high winds and storm surge topped by large waves are among the most destructive forces to be reckoned with on the coast. However, hurricanes are much less frequent than northeasters, hence the total damage caused by these lesser storms is greater. Figure 7.1 illustrates the effects of hurricane and storm forces on houses and other structures.

Winds can be evaluated in terms of the pressure they exert. Wind pressure varies with the square of the velocity, the height above the ground, and the shape of the object against which the wind is blowing. For example, a 100-mph wind exerts a pressure or force of about 40 pounds per square foot (psf) on a flat surface such as a sign. The effect on a curved surface such as a sphere or cylinder is much less—about half the force on a flat surface. If the wind picks up to 190 mph, the 40 psf force would increase to 140 psf on the flat surface.

Wind velocity increases with height above the ground, so a tall structure is subject to greater velocity and thereby greater pressure than a low structure. The velocity and corresponding pressure at 100 feet above the ground could be almost double that at ground level.

A house or building designed for inland areas is built primarily to resist vertical loads. It is assumed that the foundation and framing must support the load of the walls, floor, roof, and furniture against relatively insignificant wind forces.

A well-built house in a hurricane-prone area, however, must be constructed to withstand a variety of strong wind forces that may come from any direction. Although many people think that wind damage is caused by

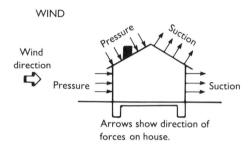

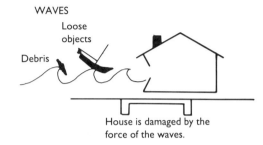

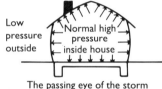

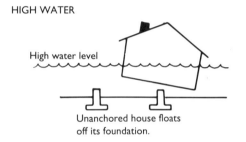

7.1 Forces to be reckoned with at the shore.

uniform horizontal pressures (lateral loads), most damage, in fact, is caused by uplift (vertical), suctional (pressure outward from the house), and torsional (twisting) forces (fig. 7.2). High horizontal pressure on the windward side is accompanied by suction on the leeward side. The roof is subject both to downward pressure and, more important, to uplift. Often a roof is sucked up by the uplift drag of the wind. Usually the failure of houses is in the devices that tie the parts of the structure together. All structural members (beams, rafters, columns) should be fastened together on the assumption that about 25 percent of the vertical load on the member may be a force coming from any direction (sideways or upward). Such structural integrity is also important if it is likely that the structure may be moved to avoid destruction by shoreline retreat.

Storm Surge and Flooding

Storm surge is a rise in sea level above the normal water level during a storm. During hurricanes and storms the coastal zone is inundated by storm surge and accompanying storm waves, and these cause most of the property damage and loss of life.

Often the wind backs water into streams or estuaries already swollen from the exceptional rainfall brought on by the hurricane or northeaster. Water is piled onto the shore by the storm. In some cases the direction of flooding may be from the bay or landward side of coastal islands. This flooding is particularly dangerous when the wind pressure keeps the tide from running out of inlets, so that the next normal high tide pushes the accumulated waters even higher. Flooding can cause an unanchored house to float off its foundation and come to rest against another house, severely damaging both.

Disaster preparedness officials have pointed out that it is a sad fact that even many condominiums built on pilings are not well-anchored or tied to those pilings. Even if the house itself is left structurally intact, flooding may destroy its contents. People who have cleaned the mud and damaged contents from a house subjected to flooding retain vivid memories of these effects.

Proper coastal development takes into account the expected level and frequency of storm surge for the area. In general, building standards require that the first habitable floor of a dwelling be above the 100-year flood level plus an allowance for wave height. At this level, a building has a 1 percent probability of being flooded in any given year.

Building or Buying a House Near the Shore **123**

Waves

Hurricane and persistent storm waves can cause severe damage not only in forcing water onshore to flood into buildings, but also in throwing boats, barges, piers, houses, and other floating debris inland against standing structures (figs. 7.1 and 7.2). The force of a wave may be understood when one considers that a cubic yard of water weighs over three-quarters of a ton; hence, a breaking wave moving shoreward at a speed of several tens of miles per hour can be one of the most destructive elements of a hurricane. Waves also can destroy coastal structures by scouring away the sand underneath them, causing them to collapse. It is possible to design buildings to survive crashing storm surf, as in the case of lighthouses, but in the balanced-risk equation it usually is not economically feasible to build ordinary cottages to resist the more powerful of such forces. On the other hand, cottages can be made considerably more storm-worthy by following the suggestions below.

Battering by Debris

The possibility of floating objects striking a house during flooding should be taken into account (fig. 7.1). To get an idea of the size of a battering load against which the house should be designed, we refer to the *Model Minimum Hurricane Resistant Building Standards for the Texas Gulf Coast* (appendix D), which specifies that the normal battering load is equal to the impact force produced by a 1,000-pound mass traveling at 10 feet per second and acting on a square foot surface of the structure.

For certain buildings, such as "Safe Refuges," the standards specify that they be constructed to resist more severe battering loads than the load listed above. The "Safe Refuge" designation refers to a building or structure located in the flood area with space sufficiently above the high water level to be authorized as a safe refuge or haven in the event of a hurricane.

Barometric Pressure Changes

Changes in barometric pressure may be a minor contributor to structural failure (fig. 7.2). If a house is sealed at a normal barometric

7.2 Modes of failure and how to deal with them. (Modified from *Wind Resistant Design Concepts for Residences* in appendix D.)

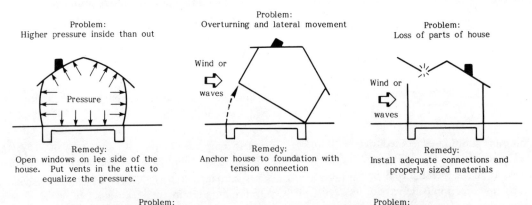
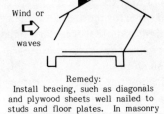
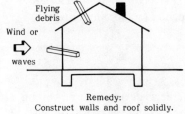

pressure of 30 inches of mercury, and the external pressure suddenly drops to 26.61 (as it did in Hurricane Camille in Mississippi in 1969), the pressure exerted within the house would be 245 pounds per square foot. An ordinary house would explode if it were leak-proof. Fortunately, houses leak air, so given the more destructive forces of storm winds and waves, pressure differential is of minor concern. Venting the underside of the roof at the eaves is a common means of equalizing internal and external pressure.

House Selection

Some types of houses are better suited to the shore than others, and an awareness of the differences will help you make a better selection, whether you're building a new house or buying an existing one.

Worst of all are unreinforced masonry houses, whether brick, concrete block, hollow clay-tile, or brick veneer, because they cannot withstand the lateral forces of wind and wave, battering by debris, flooding, or scour and settling of the foundation. If done properly, adequate and extraordinary reinforcing in coastal regions will alleviate the inherent weakness of unit masonry. Reinforced concrete and steel frames are excellent but are rarely used for small residential structures. Masonry houses also are difficult to move. Given shoreline retreat, building a movable structure may, in the long term, make more sense than protecting an immovable house.

It is hard to beat a wood-frame house that is properly braced and anchored and has well-connected members. The well-built wood house often will hold together as a unit, even if moved off its foundations, when other types would disintegrate. It is also the easiest to raise to a safer level, a common floodproofing technique for older houses. Moving the house inland as the shoreline retreats is also quite easy with a wood-frame house. In addition, the building must be designed (or modified) and adequately anchored to prevent flotation collapse, or lateral movement (figure 7.2). It must be constructed with materials and utility equipment resistant to flood damage.

Keeping Dry: Pole or Stilt Houses

In coastal regions subject to flooding by waves or storm surge, the best and most common method of minimizing damage is to raise the lowest floor of a residence above the expected water level. The first habitable floor of a home also must be above the 100-year storm-surge level (plus calculated wave height) to qualify for the National Flood Insurance Program.

7.3 Shallow and deep supports for poles and posts. (Source: Southern Pine Association.)

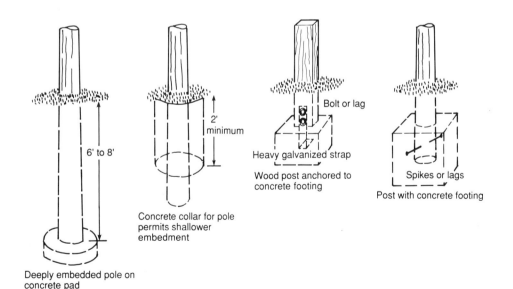

Nonresidential buildings should be floodproofed at least up to the base flood level and/or elevated at or above this level. Where the soil is suitable, most modern flood-zone structures should be constructed on pilings, well-anchored in subsoil. Elevating the structure by building a mound is adequate if flooding is the only hazard, but it is not suited to the coastal zone where mounded soil is easily eroded.

Current building design criteria for pole-house construction under the flood insurance program are outlined in the book *Elevated Residential Structures* (appendix D). Regardless of insurance, pole-type construction with deep embedment of the poles is best in areas where waves and storm surge will erode foundation material (for example, barrier islands, dune and marsh areas). Materials used in pole construction include piles, posts, and piers.

Piles are long, slender columns of wood, steel, or concrete driven into the earth to a depth sufficient to support the vertical load of the house and to withstand the horizontal forces of flowing water, wind, and waterborne debris. Pile construction is especially suitable in areas where scouring (soil washing out from under a house's foundation) is a problem.

Posts usually are wood; if steel, they are called columns. Unlike piles, a post is not driven into the ground but is placed in holes at the bottom of which may be a concrete pad (fig. 7.3). Posts may be held in place by backfilling and tamping earth or by pouring concrete into the hole after the post is in place. Posts are more readily aligned than driven piles and therefore are better to use if poles extend to the roof (as they should). In general, treated wood is the cheapest and most common material for both posts and piles.

Piers are vertical supports, thicker than piles or posts, usually made of reinforced concrete or reinforced masonry (concrete blocks or bricks). They are set on footings and extend to the underside of the floor frame.

Pole construction can be of two types. The poles can be cut off at the first-floor level to support the platform that serves as the dwelling floor. If that is what you want, piles, posts, or piers can be used. Or poles can be extended to the roof and rigidly tied into both the floor and the roof. In this way, they become major framing members for the structure and provide better anchorage to the entire house (fig. 7.4). A combination of full and floor-height poles is used in some cases, with the shorter poles restricted to supporting the floor inside the house (fig. 7.5).

Where the foundation material can be eroded by waves or winds, the poles should be deeply embedded and solidly anchored either by driving piles or by drilling deep holes for posts and putting a concrete pad at the bottom of each hole. Where the embedment is shallow, a concrete collar around the pole improves anchorage (fig. 7.3). The choice depends on soil conditions. In contrast to piles, which are more difficult to align to match the house frame, posts can be positioned in the holes before backfilling. In either case, the foundations must be deep enough to provide support after the maximum predicted loss of sand from storm erosion and scour. This required depth often dictates piles rather than posts.

Piles have the advantage of permitting far better penetration at a reasonable cost. Insufficient depth of pile or post will cause failure if storm waves liquify and erode the sand support. Improper connections of floor to piling and inadequate pile bracing will contribute to structural failure. Just as important as driving the piling deep enough to resist scouring and to support the loads they must carry is the need to fasten them securely to the structure they support. The connections must resist uplift as well as horizontal loads from wind and wave during a storm.

When post holes are dug, rather than pilings driven, the posts should extend 4 to 8 feet into the ground to provide anchorage. The post's lower end should rest on a concrete pad, spreading the load to the soil over a greater area to prevent settling. Where the soil is sandy or allows for embedment of less than about 6 feet, it is best to tie the post down to the footing with straps or other anchoring devices to prevent uplift. Driven piles should penetrate to a minimum of 8 feet. However, most soils require greater embedment, which may be required by codes for specific situations. If the site is near the water, for example, deeper embedment is needed.

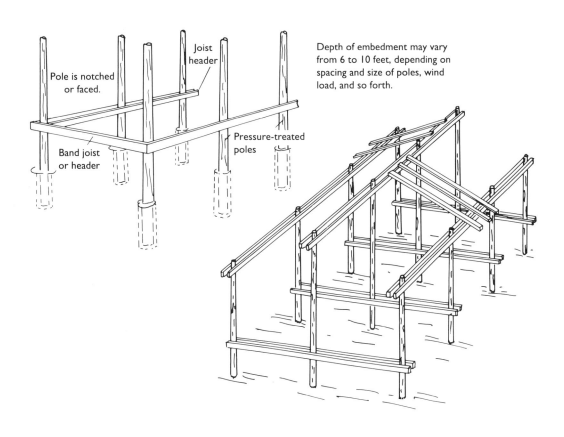

7.4 Framing system for an elevated house. (Source: Southern Pine Association.)

As mentioned, it is important to embed piles well below the depth of potential scour. Some localities require that piles be driven to a depth of at least 10 feet below mean sea level. The floor and the roof should be securely connected to the poles with bolts or other fasteners. When the floor rests on poles that do not extend to the roof, attachment is even more critical, and a system of metal straps is often used. Unfortunately, it sometimes happens that builders inadequately attach girders, beams, and joists to supporting poles by too few and undersized bolts. Hurricanes have caused greater damage than necessary as a result.

Local building codes may specify the size, quality, and spacing of the piles, ties, and bracing, as well as the methods of fastening the structure to them. Building codes often are minimal requirements, however, and building inspectors are usually amenable to allowing designs that are equally or more effective.

Piers should be constructed of precast concrete, which is stronger than concrete, and reinforced with an anchor bolt running the length of the pier. The pier should be set a minimum of 10 to 12 feet into the ground and extend 2 to 3 feet above flood level. Piers are spaced 8 to 10 feet apart.

The space under an elevated house, whether pole-type or otherwise, must be kept free of obstructions to minimize the impact of waves and floating debris. If the space is enclosed, the enclosing walls should be designed so that they

Building or Buying a House Near the Shore 127

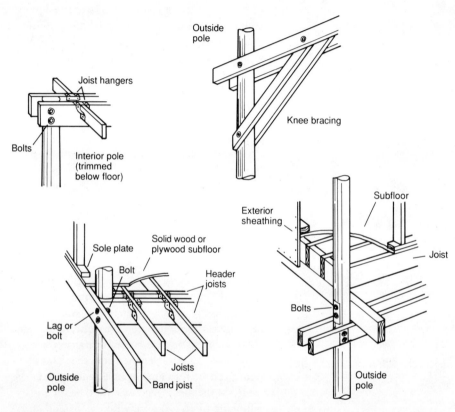

7.5 Tying floors to poles. (Source: Southern Pine Association.)

can break away or fall under flood loads, but also remain attached to the house or be heavy enough to sink. If that should happen, the walls cannot float away to add to the waterborne debris problem. An alternative is to design walls that can be swung up out of the path of the floodwaters, or else walls can be built with louvers that allow the water to pass through. The louvered wall, however, is subject to damage from floating debris. The convenience of closing in the ground floor for a garage, storage area, or recreation room may be costly because it may violate insurance requirements and actually contribute to loss of the house in a hurricane. The design of the enclosing breakaway walls should be checked against insurance requirements. See the book *Elevated Residential Structures* (appendix D).

Recommendations for building design, framing, and construction materials will be found in the following section.

An Existing House: What to Look For, What to Improve

If instead of building a new house, you select a house already built in an area subject to waves, flooding, and high winds, consider (1) where the house is located; (2) how well the house is built; and (3) how the house can be improved.

Geographic Location

Evaluate the site of an existing house using the same principles given earlier for evaluating a possible site for a new house. House elevation, frequency of high water, escape routes, and how well the lot drains should be foremost, but you should go through the complete site-safety checklist.

A house can be modified after you purchase it, but you can't prevent hurricanes or other storms. The first step is to stop and consider: do the pleasure and benefits of this location balance the risk and disadvantages? If not, look elsewhere; if so, then evaluate the house itself.

How Well Is the House Built?

In general, the principles used to evaluate an existing house match those used in building a new one. Remember, though, that many houses predate enactment of the National Flood Insurance Program and may not meet the standards or improvements required of structures built since then.

Before you thoroughly inspect the building in which you're interested, look closely at the adjacent structures. If poorly built, they may float over against your building and damage it in a flood. You may even want to consider the type of people you will have as neighbors: will they clear the decks to prepare for a storm or will they leave items in the yard to become windborne missiles?

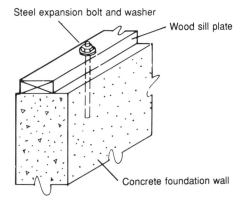

7.6 Additional connection of wood-frame building to foundation. Temporarily remove the wall covering enough to add half-inch or larger steel expansion bolts to gain additional anchorage.

Ground Anchorage and Foundation

The house or condominium itself should be inspected to see that the structure is well-anchored to the ground. If it simply rests on blocks, rising water may cause it to float off its foundation and come to rest against your neighbor's house or in the middle of the street. In New Jersey the 1962 Ash Wednesday storm carried many homes all the way into back bay waters. If that should happen, it may be possible to move a well-built, internally well-braced house back to its proper location, but chances are great that the structure will be too severely damaged to be habitable.

If the building is on piles, posts, or poles, check to see if the floor beams are adequately bolted to them. If it rests on piers, crawl under the house to see if the floor beams are securely connected to the foundation. If the floor system rests unanchored on piers, *don't* buy the house.

It is difficult to discern whether a building built on a concrete slab is properly bolted to the slab because the inside and outside walls hide the bolts. But it may be possible to ascertain if the wall is bolted to the slab by stripping away some of the bottom part of the inside wall and exposing the sill or plate and portions of the anchor bolts. If the building frame (walls and floor) is not anchored to the slab or to the foundation wall, it may be possible to achieve this anchorage by temporarily removing enough wall covering to add expansion bolts (fig. 7.6) of a half-inch (1.3 cm) or larger.

Suppose your house is resting on blocks but not fastened to them and is thus not adequately anchored to the ground. Can anything be done? Yes. Several solutions are possible. Perhaps the house's configuration will allow piles to be driven deep into the ground at each corner and then fastened to the house. This method of anchoring can apply to mobile homes as well as houses. (See the section on mobile homes.)

Diagonal struts under the house—either timber or pipe—may also anchor a house that rests on blocks. This is done by fastening the upper ends of the struts to the floor system and the lower ends to individual concrete footings substantially below ground. These struts must be able to sustain both uplift (tension) and compression and should be tied into the concrete footing with anchoring devices such as metal straps or spikes.

If you can locate the builder, ask if such bolting was done. Better yet, if you can get assurance that construction of the house complied with the provisions of a building code serving the special needs of a coastal region, you can be reasonably sure that all parts of the house are well-anchored: the foundation to the ground, the floor to the foundation, the walls to the floor, and the roof to the walls (figs. 7.7, 7.8, and 7.9). Be aware that many builders, carpenters, and building inspectors who are accustomed to traditional construction are apt to regard metal connectors, collar beams, and

7.7 Foundation anchorage. Top: anchored sill for shallow embedment. Bottom: anchoring sill or plate to foundation. (Bottom modified from *Houses Can Resist Hurricanes* in appendix D.)

7.8 Stud-to-floor, plate-to-floor framing methods. (Modified from *Houses Can Resist Hurricanes* in appendix D.)

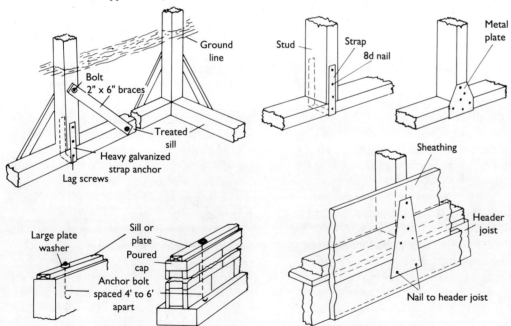

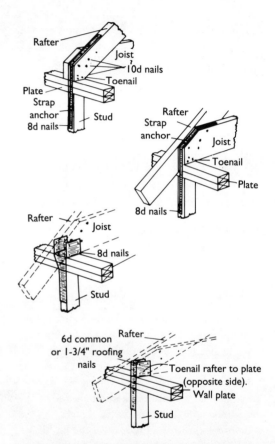

7.9 Roof-to-wall connectors. The top drawings show metal strap connectors. Left, rafter to stud; right, joist to stud. The bottom left drawing shows a double-member metal plate connector—in this case with the joist to the right of the rafter. The bottom right drawing shows a single-member metal plate connector. (Modified from *Houses Can Resist Hurricanes* in appendix D.)

other such devices as newfangled and unnecessary. If consulted, they may assure you that a house is as solid as a rock, when, in fact, it is far from it. Nevertheless, it's wise to consult the builder or knowledgeable neighbors when possible.

Roof
The roof should be well-anchored to the walls (fig. 7.9); this will prevent uplifting and separation from the walls. Visit the attic to see if such anchoring exists. Simple toe-nailing (nailing at an angle) is not adequate; metal fasteners are needed. Depending on the type of construction and the amount of insulation laid on the attic floor, these fasteners may or may not be easy to see. If roof trusses or braced rafters were used, it should be easy to see whether the various members, such as the diagonals, are well-fastened together. Again, simple toe-nailing will not suffice. Some builders, unfortunately, nail parts of a roof truss just enough to hold it together to get it in place. A collar beam or gusset at the peak of the roof (fig. 7.10) provides some assurance of good construction. The Standard Building Code states that wood rafters shall be securely fastened to the exterior walls with approved hurricane anchors or clips.

Be sure to look at the condition and composition of the roof. Quality roofing material should be well-anchored to the sheathing. A poor roof covering will be destroyed by hurricane-force winds, allowing rain to enter the house and damage ceilings, walls, and contents.

Wood shingles and shakes, properly fastened, resist storm damage better than most roofing materials. For best performance in areas subject to storms, shakes and shingles should be connected to wood sheathing by two galvanized nails per shingle, long enough to penetrate the sheathing. If the sheathing is plywood, the nails should be threaded. For roof slopes that rise 1 foot or more for every 3 feet of horizontal distance, exposure of the shingles should be about one-fourth of each one's length—4 inches (10 cm) for a 16-inch-long (41 cm) shingle. If shakes (thicker and longer than shingles) are used, less than one-third of their length should be exposed—about 6.5 to 7 inches (16 to 18 cm) for a 24-inch-long (61 cm) shake.

In hurricane areas asphalt shingles should be

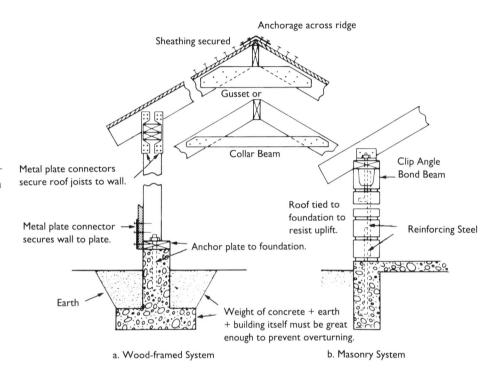

7.10 Where to strengthen a house.

Building or Buying a House Near the Shore

exposed somewhat less than is usual inland. A mastic or seal-tab type or an interlocking shingle of heavy grade is recommended. Roof underlay of asphalt-saturated felt should be used with six galvanized roofing nails or approved staples for each three-tab strip in a square-butt shingle. For low-pitch roofs, double coverage of the underlayment is preferred.

On built-up roofing that consists of layers of asphalt-saturated felt, the surfacing aggregate should be fully imbedded in the surface coating to minimize damage by flying gravel to adjacent windows, cars, and painted objects during high winds.

Corrugated metal and asbestos cement sheets are satisfactory if held down properly to the roof structure so they cannot detach and become missiles. If a wood deck (closely spaced boards or plywood) is used, these sheets could be secured with drive screws of sufficient length to extend through the deck. If the sheets rest directly on purlins or other roof members, they should be secured with strap fasteners, bolts, or stud fasteners, or by properly designed clip fasteners. Since Cyclone Tracy in Darwin, new construction uses battens or strips on top of roof sheets to aid in holding them down in a high wind.

If the roofing sheets are aluminum, avoid iron nails or screws; the two materials are not compatible, and combining them results in corrosion. Aluminum nails are available and should be used. Likewise, aluminum roofing should be insulated when fastened to a steel-roof structure to prevent electrogalvanic action.

Framing and Design

The fundamental rule to remember in framing is that all structural elements should be fastened together and anchored to the ground in such a way as to resist all forces, regardless of which direction they come from. This principle of sturdy connections prevents overturning, floating off, racking, or disintegration.

If the house has a porch with exposed columns or posts, it should be possible to install tie-down anchors on their tops and bottoms. Steel straps or clips should suffice in most cases.

If there is an addition to the house, be sure that the new construction is adequately fastened to the existing structure. Such additions are particularly vulnerable to detachment and damage during storms.

Where accessible, roof rafters and trusses should be anchored to the wall system. On a completed house, the juncture of roof and wall is often extremely difficult to reach. Except where they meet the walls, the roof trusses or braced rafters are usually sufficiently exposed to make it possible to strengthen joints (where two or more members meet)—particularly at the peak of the roof—with collar beams or gussets (fig. 7.10).

A competent carpenter, architect, or structural engineer can review the house with you and help decide what modifications are most practical and effective. Do not be misled by someone who resists new ideas. One builder actually told a homeowner, "You don't want all those newfangled straps and anchoring devices. If you use them the whole house will blow away, but if you build in the usual manner [with members lightly connected] you may lose only part of it."

In fact, the very purpose of the straps is to prevent any or all of the house from blowing away. As the Standard Building Code states: "Lateral support securely anchored to all walls provides the best and only sound structural stability against horizontal thrusts, such as winds of exceptional velocity."

Consider bracing or strengthening the interior walls where feasible. Such reinforcement may require removing the surface covering and installing as much plywood sheathing or strap bracing as possible. Where wall studs are exposed, bracing straps offer a simple way to achieve needed reinforcement against the wind. These straps are commercially produced and are made of 16-gauge galvanized metal with prepunched holes for nailing. These should be secured to studs and wall plates as nail holes permit. A 10d (tenpenny) nail is 3 inches long, and an 8d, 2.5 inches long. Where compression forces are to be resisted, use 1 by 6 lumber with three 8d or 10d nails at 8-inch centers. Bracing is important not only along the length of the wall, but it should be provided at right angles to the wall's loaded

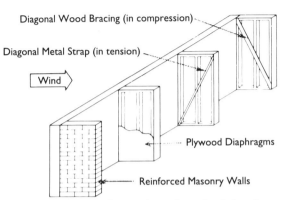

7.11 Bracing walls at right angles to loaded surface. (From Defense Civil Preparedness Agency, Publication TR-83.)

above). The pressure exerted by a wind on a round or elliptical shape is about 60 percent of that exerted on the common square or rectangular shape; the pressure exerted on a hexagonal or octagonal cross section is about 80 percent of that exerted on a square or rectangular cross section (fig. 7.12).

The design of a house or building in a coastal area should minimize structural discontinuities and irregularities. It should be plain and simple and have a minimum of nooks and crannies and offsets on the exterior surface at about 12-foot centers to the extent possible. Remember that the wind force may come from any direction. Figure 7.11 illustrates bracing at right angles to a wall.

The shape of the house is important. A four-sided, steep-pitched hip roof can resist high winds better than a low-pitched gable roof, which slopes in only two directions. This was found to be true in Hurricane Camille in 1969 in Mississippi, in Typhoon Tracy, which devastated Darwin in 1974, and in Hurricane Hugo. The reason is twofold: the hip roof offers a smaller shape for the wind to blow against, and its structure is such that the roof is better braced in all directions. Low-pitched roofs as well as overhangs and porches add to the wind's lift potential.

Note also the horizontal cross section of the house (the shape of the house as viewed from

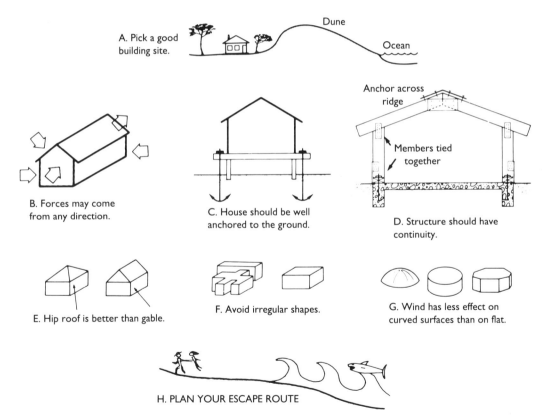

7.12 Some rules in selecting or designing a house.

Building or Buying a House Near the Shore

because damage to a structure tends to concentrate at these points. Some of the newer beach cottages are of a highly angular design with such nooks and crannies. Award-winning architecture will be a storm loser if the design has not incorporated the technology for maximizing structural integrity in the face of storm forces. In the absence of irregularities, the house reacts to storm winds as a complete unit (fig. 7.12).

Will the house leak enough air to combat changes in barometric pressure? If the house has an overhanging eave and no openings on its underside, it may be feasible to cut openings and screen them. These openings keep the attic cooler (a plus in the summer) and help to equalize the pressure inside and outside during a storm with a low-pressure center.

Brick, Concrete, and Masonry Reinforcement

Brick, concrete-block, and masonry-wall houses should be adequately reinforced. This reinforcement is hidden from view. Building codes applicable to high-wind areas often specify the type of mortar, reinforcing, and anchoring to be used in construction. If you can get assurance that the house was built in compliance with a building code designed for such an area, consider buying it. *At all costs, avoid unreinforced masonry houses.*

A poured concrete bond beam at the top of the wall just under the roof is one indication that a house is well-built (fig. 7.13). Most

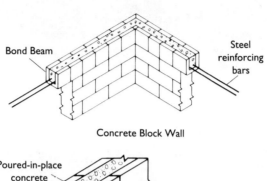
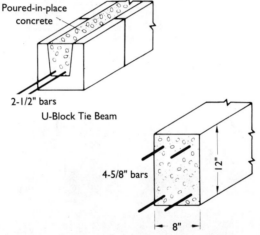

7.13 Reinforced tie beam (bond beam) for concrete block walls, to be used at each floor level and at roof level around the perimeter of the exterior walls.

bond beams are formed by putting in reinforcing and pouring concrete into U-shaped concrete blocks. From the outside, however, since these U-shaped blocks are indistinguishable from ordinary ones, you cannot be certain that the bond beam exists. The vertical reinforcing should penetrate the bond beam.

Some architects and builders use a stacked bond (one block directly above another), rather than overlapped or staggered blocks, because they believe it looks better. The stacked bond is definitely weaker than the staggered block. Unless you have proof that the walls are adequately reinforced to overcome this lack of strength, avoid this type of construction.

In past hurricanes the brick veneer of many houses has separated from the wood frame, even when the houses remained standing. Asbestos-type outer wall panels used on many houses in Darwin were found to be brittle, and they broke up under the impact of windborne debris in Typhoon Tracy. Both types of construction—brick veneer and asbestos-type panel—should be avoided along the coast.

Glass Surfaces

Glazing (windows, glass doors, glass panels) should be kept to a minimum. Although large open glass areas facing the open water provide excellent bay or sea views, such glazing presents an obvious hazard. Glass disintegrates and blows inward during a storm, and such glass projectiles are lethal. Less frequently recognized problems are these: glass may not provide as much structural strength as wood, metal, or other building materials; and ocean-facing glass is commonly damaged through sand blasting by sediment that is transported by normal coastal winds. Sand blasting may be avoided by reducing the amount of glass in the original design, or storm shutters made of materials from steel to wood can be installed.

In-House Shelter

If you wish to improve the protective capability of your house, several methods are available even if you own an existing house and are limited to in-place construction modifications.

(1) You can strengthen the entire house. While this is easier to do as the house is being built, much can be accomplished even with an existing structure.

(2) If there are budget constraints or you don't want to change the house's appearance, you may choose to strengthen only part of the structure.

(3) A shelter module can be built within the house for refuge during a storm. The advantages of such an in-house shelter module are that it is quickly accessible, has a daily usefulness, and its protective features can be visually and functionally blended to fit the residence. To choose where the shelter will be located, examine the house and select the best room to stay in during a wind storm. Note that we emphasize *wind storm*. Such a shelter is not an alternative to evacuation prior to a hurricane! A small windowless room, such as a bathroom, utility room, den, hall, storage or mechanical equipment/laundry room is usually stronger than one with windows (fig. 7.14). A sturdy inner room with more than one wall between it and the outside is safest. The fewer the doors, the better; an adjoining wall or baffle or obstructing wall shielding the door adds to the protection. The forces to be resisted are pressure and suction from the velocity of the wind.

To summarize, a shore house should have (1) the roof tied to walls, walls tied to the foundation, and the foundation anchored to the earth (the connections are potentially the weakest link in the structural system); (2) a shape that resists storm forces; (3) floors high enough (of sufficient elevation) to be above most storm waters (usually the 100-year flood level plus 3 to 8 feet); (4) piles or posts that are of sufficient depth or embedded in concrete to anchor the structure and withstand erosion; and (5) well-braced piling. Consult a good architect or structural engineer for advice if you have doubts or questions. A few dollars for wise counsel may save you from later financial grief.

Summary: Protective Steps for Strengthening Buildings

During Hurricane Hugo, well-maintained houses in Charleston, S.C., were found to have suffered less damage than houses that were not well cared for. Go over your house carefully and find anything that needs to be repaired—and do it right away.

(1) Are there openings in the attic under the eaves? If not, put them in.

(2) To see if the house is bolted to the slab, remove some of the lower inside wall and look to see if there *are* bolts. If the walls need strengthening, add diagonals or plywood to the walls.

(3) Strengthen existing poles by excavating around the pole and backfilling with concrete.

(4) You can replace big windows with smaller ones or with stronger glass. Check all windows to be sure they are well-fastened to the walls. If storm shutters are lacking, add them for additional protection.

(5) To prevent the house from sliding off foundation blocks, use an anchor system for mobile homes, or drive posts at the corners of

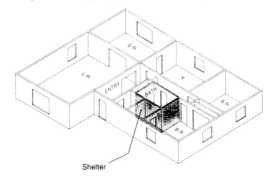

7.14 Bathroom shelter module. (From Defense Civil Preparedness Agency, Publication TR-83.)

the house and tie them into the structure.

(6) On an elevated house, add knee braces to poles and fasten to the floor system.

(7) Where applicable, add collar beams to roof trusses.

(8) Examine every visible bolted joint, and, if they are not well-fastened, add more bolts.

(9) Examine the roof. If there are not enough nails in the roof, add them; or if the shingles are not overlapped, add a new row and be sure the shingles are nailed to the sheathing battens with galvanized nails. If you need more protection, add battens or strips on top of the roof to hold the shingles down.

(10) Check and improve the fastening of the roof to the house frame (walls).

(11) Tie down the porch to the house and the roof of the porch to the posts.

If you have any other additions to the house, be aware that these are weak points to wind forces and should be securely tied to the house.

Mobile Homes: Limiting Their Mobility

Because of their light weight and flat sides, mobile homes are particularly vulnerable to the high winds of hurricanes, tornadoes, and severe storms. Such winds can overturn unanchored mobile homes or smash them into neighboring homes and property. Millions of Americans live in mobile homes today, and the number is growing. Twenty to 30 percent of single-family housing production in the United States consists of mobile homes. High winds damage or destroy nearly 5,000 of these homes every year, and the number will surely rise unless greater protective measures are taken. As one man whose mobile home was overturned in Hurricane Frederic (1979) put it, "People who live in flimsy houses shouldn't have hurricanes."

Several lessons can be learned from past storm experiences. First, mobile homes should be properly located. After Hurricane Camille (1969), it was observed that where mobile-home parks were surrounded by woods and where units were close together, minimal damage was caused, mainly by falling trees. In unprotected areas, however, many mobile homes were overturned and often destroyed from the force of the wind. The protection afforded by trees is greater than the possible damage from falling limbs. Two or more rows of trees are better than a single row, and trees 30 feet or more in height give better protection than shorter ones. If possible, mobile homes should be positioned so that the narrow side faces the prevailing winds.

Locating a mobile home in a hilltop park will greatly increase its vulnerability to the wind. A lower site screened by trees is safer from the wind, but it should be above storm-surge flood levels. A location that is too low obviously increases the likelihood of flooding. There are fewer safe locations for mobile homes than for stilt houses, although mobile homes can be elevated on pilings similar to pole or stilt houses, as has been done at the Chimney Creek subdivision near Tybee Island.

A second lesson taught by past experience is that the mobile home must be tied down or anchored to the ground so that it will not overturn in high winds (figs. 7.15, 7.16; table 7.1). Simple prudence dictates the use of tiedowns. Many insurance companies, moreover, will not insure mobile homes unless they are adequately anchored with tiedowns. A mobile home may be tied down with cable or rope, or rigidly attached to the ground by connecting it to a simple wood-post foundation system. A conscientious mobile-home park owner can provide permanent concrete anchors or piers to which hold-down ties may be fastened. In general, an entire *tiedown* system costs only a nominal amount.

A mobile home should be properly anchored with ties to the frame and over-the-top straps; otherwise, the home may be damaged by sliding, overturning, or tossing. The most common cause of major damage is the tearing away of most or all of the roof. When this happens, the walls are no longer adequately supported at the top and are more prone to collapse. Total destruction of a mobile home is more likely if the roof blows off, especially if the roof blows off first and then the home overturns. The need for anchoring cannot be overemphasized; there should be over-the-top tiedowns to resist overturning, and frame ties to resist sliding off the piers. This applies to single mobile homes up to 14 feet in width. Double-wides do not require over-the-top ties, but they do require frame ties. Although newer mobile homes are equipped with built-in straps to aid in tying down, the occupant may wish to add more if the home is in a particularly vulnerable location. Many older mobile homes are not equipped with these built-in straps.

Also keep in mind that most tiedown system requirements are designed to withstand 70–110 mph winds. Since hurricane winds can reach speeds of 190 mph, it is imperative that you *always* evacuate a mobile home in the event of a hurricane.

For more information, obtain a copy of *Manufactured Home Installation in Flood Hazard Areas* from FEMA (appendix D). This reference treats the subject in more detail.

High-Rise Buildings: The Urban Shore

A high-rise building on the beach is generally designed by an architect and a structural engineer who are presumably well-qualified and aware of building requirements on the shoreline. Building tenants, however, should not assume that the structure is invulnerable. Residents of 2- and 3-story apartment buildings were killed when the buildings were destroyed by Hurricane Camille in Mississippi in 1969. In Delaware storms have smashed five-story buildings, and larger high-rises have yet to be thoroughly tested by a major hurricane.

The most important aspect of high-rise construction that a prospective dweller or condo owner must consider is the quality of the building's foundation. High-rises near the beach should be built so that even if the foundation is severely undercut during a storm the building remains standing. Shortcuts are sometimes taken by less scrupulous builders, and piling is not driven deeply enough. Just as important is the need to fasten piles securely to the structure they support. The connections must resist horizontal loads from wind and wave during a storm as well as uplift. Builders and building inspectors are jointly responsible for making sure the job is done right. The foundation of a high-rise under construction at Panama City Beach, Florida, was exposed by Hurricane Eloise to reveal that 30 of its pilings had no concrete around them and were not attached to the building. Such problems exist wherever high-rises crowd the beach.

Despite assurances that come with an engineered structure, life in a high-rise building holds definite drawbacks that prospective tenants should consider. These conditions stem from high winds, high water, and poor foundations.

Wind pressure is greater near the shore than it is inland, and it increases with a structure's height. If you live inland in a two-story house and move to the eleventh floor of a high-rise on the shore, you should expect five times more wind pressure than you're accustomed to. This can be a great and possibly devastating surprise.

High wind pressure can actually cause

Table 7.1 Tiedown Anchorage Requirements

	10- and 12-foot wide mobile homes				12- and 14-foot wide mobile homes	
	30 to 50 feet long		50 to 60 feet long		60 to 70 feet long	
Wind velocity (mph)	Number of frame ties	Number of over-the-top ties	Number of frame ties	Number of over-the-top ties	Number of frame ties	Number of over-the-top ties
70	3	2	4	2	4	2
80	4	3	5	3	5	3
90	5	4	6	4	7	4
100	6	5	7	5	8	6
110	7	6	9	6	10	7

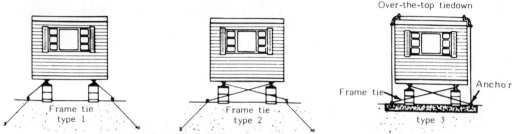

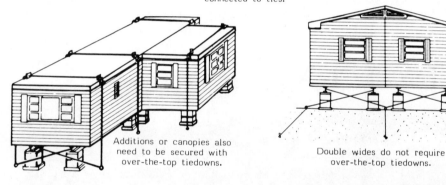

7.15 Tiedowns for mobile homes. (Modified from *Protecting Mobile Homes from High Winds* in appendix D.)

motion sickness because of the building's unpleasant swaying. It is worthwhile to check with current residents to find out if a high-rise has these undesirable characteristics. More seriously, high winds can break windows, damage property, and inflict injury. Remember, too, that tenants of severely damaged buildings will have to relocate until repairs are made, and this can have serious consequences for some people.

Those interested in researching the subject further—even the knowledgeable engineer or architect who is hired to design a shore structure—should obtain a copy of *Structural Failures: Modes, Causes, Responsibilities* (see appendix D); of particular importance is the chapter "Failure of Structures Due to Extreme Winds," which analyzes wind damage to engineered high-rise buildings from the storms at Lubbock and Corpus Christi, Texas, in 1970.

Power failures or blackouts also affect a multifamily high-rise building more seriously than a low-occupancy structure. These are more likely along the shore than inland because of the more severe weather conditions associated with coastal storms. A power failure can cause great distress. People can be caught between floors in elevators. On a large scale, New York City once had such an experience. Think of the mental and physical distress after several hours of confinement, and compound this with the roaring winds of a storm whipping around the building, sounding like a runaway freight train. In this age of electricity it is easy to imagine many other inconveniences that can be caused by power failures in multistory buildings.

Fire is an additional hazard in a high-rise building. Even recently constructed buildings seem to have such difficulties. (The television pictures of a woman leaping from the window of a burning building in New Orleans to avoid being incinerated are a horrible reminder from recent history.) The many hotel fires over the past few years demonstrate the problems. Fire Department equipment reaches only so high. And many areas along the coast are too sparsely populated to afford high-reaching equipment. Inadequate water pressure may also be a problem in some communities.

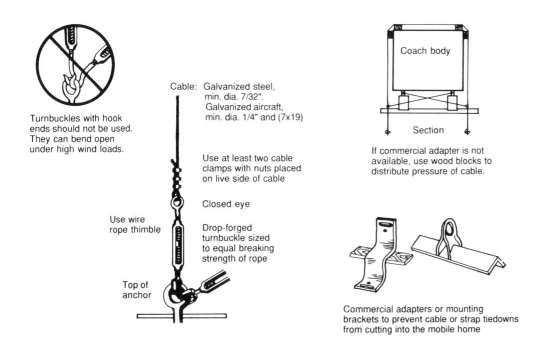

7.16 Hardware for mobile home tiedowns. (Modified from *Protecting Mobile Homes from High Winds* in appendix D.)

Fire and smoke travel along ventilation ducts, elevator shafts, corridors, and similar passages. But the potential for such hazards can be reduced and the building made safer, especially if it is new. Sprinkler systems should be operated by gravity water systems rather than by powered pumps (because of possible power failures). (Gravity systems use water from tanks higher up in the building.) Battery-operated emergency lights that come on only when other lights fail, better fire walls and automatic sealing doors, pressurized stairwells, and emergency-operated elevators in pressurized shafts, all will contribute to greater safety. Unfortunately, all of these improvements cost money and are often omitted unless required by the building code.

Modular Unit Construction: Prefabricating the Urban Shore

The method of building a house, duplex, or large condominium structure by fabricating modular units in a shop and assembling them at the site is gaining in popularity for construction on shoreline property. The larger of these structures are commonly two or three stories in height, and they may contain a large number of living units.

Modular construction makes good economic sense, and there is nothing inherently wrong in this approach. These methods have been used for years in manufacturing mobile homes, although final assembly is done in the shop rather than in the field. Doing as much of the work as possible in a shop can save considerably in labor and costs. Workers are not affected by outside weather conditions; they often can be paid by piecework, enhancing their productivity; shopwork lends itself to labor-saving equipment such as pneumatic nailing guns and overhead cranes.

If the manufacturer desires it, shop fabrication can permit higher quality. Inspection and control of the whole process is much easier. For instance, there is less hesitation about rejecting a poor piece of lumber when you have a nearby supply than if you're building a structure and have just so much lumber on the site.

On the other hand, because so much work is done out of the buyer's sight, the manufacturer can take shortcuts if so inclined. It is possible

Building or Buying a House Near the Shore

that some modular dwelling units have their wiring, plumbing, ventilation, and heating and air conditioning installed at the factory by unqualified personnel, and it is possible that the resulting inferior work is either not inspected at all or inspected by an unconscientious, inept, or overworked and underpaid individual. Therefore, it is important to consider the following: were wiring, plumbing, heating and air conditioning, and ventilation installed in the factory or at the building site? Were the installers licensed and certified? Was the work inspected both at the factory and on the construction site? Most important, is the modular dwelling unit built to provide safety in the event of fire? For example, just a few of the many safety features that should be included are two or more exits, stairs remote from each other, masonry fire walls between units, noncombustible wall sheeting, and compartmentalized units that prevent fire spreading.

Further, if one unit is placed on top of another, it is vital that they be adequately fastened together to resist high winds; they should not depend solely on the weight of the upper unit to hold both units in place.

In general, it is highly desirable to check a manufacturer's reputation and integrity, just as you would when hiring a contractor to build your house on-site. In other words, you should use the same caution in buying a modular unit that you would use with other structures.

As with all other types of structures, consider site safety and escape routes from the building and from the area.

Living with Nature: Prudence Pays

Hurricane or calm, receding shore or accreting shore, storm-surge flood or sunny sky, migrating dune or maritime forest, win or lose, the gamble of coastal development will continue. If you choose your site with natural safety in mind, follow structural engineering design principles in construction, and take a generally prudent approach to living at the shore (fig. 7.12), you increase the potential for both enjoying the shore and protecting your investment by reducing the risk.

Our goal is to provide guidance to today's and tomorrow's coastal dwellers. This book is not the last or by any means the complete guide to coastal living, but it should provide a beginning. In the appendixes that follow are additional resources that we hope every reader will pursue.

Appendix A   Hurricane Checklist

Keep this checklist handy for protection of family and property.

When (or Before) a Hurricane Threatens

—Most important, *know the official evacuation route* for your area. Maps of hurricane evacuation routes are published in your phone book. You will not be asked to evacuate unless your life is in danger, so *evacuate as directed* by local emergency preparedness officials.
—Read your newspaper and listen to radio and television for official weather reports and announcements.
—Secure reentry permits if necessary. Some communities allow only property owners and residents with proper identification or tags to return in the storm's immediate wake.
—Pregnant women, the ill, and the infirm should call a physician for advice.
—Be prepared to turn off gas, water, and electricity where it enters your home.
—Make sure your car's gas tank is full.
—Secure your boat. Use long lines to allow for rising water.
—Secure movable objects on your property:
  —doors and gates
  —outdoor furniture
  —shutters
  —garden tools
  —hoses
  —garbage cans
  —bicycles or large sports equipment
  —barbecues or grills
  —other
—Board up or tape windows and glassed areas. Close storm shutters. Draw drapes and window blinds across windows and glass doors, and remove furniture in their vicinity.
—Check mobile home tie-downs.
—As emphasized, your primary line of defense is *early evacuation*. In the event that you are unable to evacuate, you should also:
—Know the location of the nearest emergency shelter. Go there if directed by emergency preparedness officials.
—Fill tubs and containers with water (a minimum of one quart per person per day).
—Stock adequate supplies:
  —transistor (or battery-powered) radio
  —fresh batteries
  —canned heat
  —hammer
  —boards
  —pliers
  —hunting knife
  —tape
  —first-aid kit
  —prescribed medicines
  —candles
  —matches
  —nails
  —screwdriver
  —ax*
  —rope*
  —plastic drop cloths, waterproof bags, material for tying
  —containers of water
  —canned food, juices, soft drinks**
  —water purification tablets
  —insect repellent
  —gum, candy
  —life jackets
  —charcoal bucket, charcoal, and charcoal lighter
  —buckets of sand
  —disinfectant
  —flashlights
  —hard-top headgear
  —fire extinguisher
  —can openers and utensils (knives, forks, spoons, cups)

Special Precautions for Apartments/ Condominiums

—Designate one person as the building captain to supervise storm preparations.
—Know your exits.
—Count the stairs to exits since you may be evacuating in darkness.

---

*Take an ax (to cut an emergency escape opening) if you go to the upper floors or attic of your home. Take rope for escape to ground when water subsides.
**Stock enough food for at least three days and enough water for more than three days. Select food that does not require cooking or refrigeration. Contact the Coastal Georgia Regional Development Center for a complete three-day shopping list.

—Locate safest areas for occupants to congregate.
—Close, lock, and tape windows.
—Remove loose items from terraces (and from your absent neighbor's terraces).
—Remove or tie down loose objects from balconies or porches.
—Assume other trapped people may wish to use the building for shelter.

Special Precautions for Mobile Homes

—Pack breakables in padded cartons and place on floor.
—Remove bulbs, lamps, mirrors, and put them in the bathtub.
—Tape windows.
—Turn off water, propane gas, electricity.
—Disconnect sewer and water lines.
—Remove awnings.
—*Leave*. (Don't stay inside for *any* reason.)

Special Precautions for Businesses

—Take photographs of building and merchandise before the storm.
—Assemble insurance policies.
—Move merchandise away from plate glass.
—Move merchandise to the highest locations possible.
—Cover merchandise with tarps or plastic.
—Remove outside display racks and loose signs.

—Take out lower file drawers, wrap in trash bags, and store in high places.
—Sandbag spaces that may leak.
—Take special precautions with reactive or toxic chemicals.

If You Remain at Home

—*Never* remain in a mobile home; seek official shelter.
—Stay indoors. Remember, the first calm may be the hurricane's eye. Remain indoors until an official all-clear is given.
—Stay on the *downwind* side of the house. Change your position as the wind changes.
—If your house has an inside room (away from all outdoor walls), it may be the most secure part of the structure. Stay there.
—Keep a continuous watch for *official* information on radio and television.
—Keep calm. Your ability to meet emergencies will help others.

If Evacuation Is Advised

—Leave as soon as you can. Follow official instructions only. Ignore rumors.
—Follow predesignated evacuation routes unless those in authority direct you to do otherwise.
—Take these supplies:
  —re-entry permit

  —change of warm, protective clothes
  —first-aid kit
  —baby formula
  —identification tags: include name, address, and next of kin (wear them)
  —flashlight
  —food, water, gum, candy
  —rope, hunting knife
  —waterproof bags and ties
  —can opener and utensils
  —disposable diapers
  —special medicine
  —blankets and pillows in waterproof casings
  —transistor (or battery-powered) radio
  —fresh batteries (for radio *and* flashlight)
  —bottled water
  —purse, wallet, valuables
  —life jackets
  —games and amusements for children
—Disconnect all electric appliances except refrigerator and freezer. Their controls should be turned to the coldest setting and the doors kept closed.
—Leave food and water for pets. Seeing-eye dogs are the only animals allowed in the shelters.
—Shut off water at the main valve (where it enters your home).
—Lock windows and doors.
—Keep important papers with you:
  —driver's license and other identification
  —insurance policies

—property inventory
—Medic Alert or other advice to convey special medical information

During the Hurricane

—Stay indoors and away from windows and glassed areas.
—If you're advised to evacuate, *do so at once.*
—Listen for continuing weather bulletins and official reports.
—Use your telephone only in an emergency.
—Follow official instructions only. Ignore rumors.
—Keep a window or door *open* on the side of the house opposite the storm winds.
—Beware the eye of the hurricane. A lull in the winds does not necessarily mean that the storm has passed. Remain indoors unless emergency repairs are necessary. Be cautious. Winds may resume suddenly, in the opposite direction and with greater force than before. Remember, if wind direction does change, the open window or door must be changed accordingly.
—Be alert for rising water.
—If electric service is interrupted, note the time.
  —Turn off major appliances, especially air conditioners.
  —Do not disconnect refrigerators or freezers. Their controls should be turned to the coldest setting and doors closed to preserve food for as long as possible.
  —Keep away from fallen wires. Report location of such wires to the utility company.

If You Detect Gas

—Do not light matches or turn on electrical equipment.
—Extinguish all flames.
—Shut off gas supply at the meter.*
—Report gas service interruptions to the gas company.

Water

—The only *safe* water is the water you stored before it had a chance to come in contact with floodwaters.
—Should you require an additional supply, be sure to boil water for 30 minutes before using.
—If you're unable to boil water, treat water you will need with water purification tablets.
Note: An official announcement will proclaim tap water safe. Boil or treat all water except stored water until you hear the announcement.

*Gas should be turned back on by only a gas serviceman or licensed plumber.

After the Hurricane Has Passed

—Listen for official word of danger having passed. Don't return to your home until officially directed to do so.
—Watch out for loose or hanging power lines as well as gas leaks. People have survived storms only to be electrocuted or burned. Fire protection may be nonexistent because of broken power lines.
—Walk or drive carefully through storm-damaged areas. Streets will be dangerous because of debris, undermining by washout, and weakened bridges. Watch for poisonous snakes and insects driven out by floodwaters.
—Eat nothing and drink nothing that has been touched by floodwaters.
—Place spoiled food in plastic bags and tie securely.
—Dispose of all mattresses, pillows, and cushions that have been in floodwaters.
—Contact relatives as soon as possible.
Note: If you are stranded, signal for help by waving a flashlight at night or white cloth during the day.

Appendix B   Recent Hurricanes in Georgia

Since most current coastal residents have never experienced a major hurricane, we must look to historical accounts to gain some appreciation of the awesome power of these "September gales." In days gone by, our only knowledge of a hurricane's strength came from sketchy newspaper accounts describing the damages and number of lives lost or estimating the losses in dollars. Major storms such as those in 1804, 1854, 1881, 1893, and 1898 were all characterized as "the worst ever" or "greater than" some previous "worst" storms. The hurricanes were not, in fact, increasing in intensity, but each storm followed a somewhat different path: the full fury would be felt by Savannah in one storm and at Brunswick or Darien in the next. In recent times, storms have been compared in terms of dollar losses, but these figures primarily reflect the nature of the development that has been struck rather than the hurricane's strength.

The story of the "Antigua-Charleston hurricane" of 1804 was recounted eloquently by Aaron Burr, who had retired to a St. Simons Island plantation after his historic duel with Alexander Hamilton. To his daughter, Theodosia, he wrote of his experience as the storm passed inland. Of the 7-foot storm surge, he noted that the high waters were "sufficient to inundate a great part of the coast, to destroy all the rice, to carry off most of the buildings that were in the lowlands and destroy the lives of many. . . ."

Damage to the north was also extensive. In Savannah "the water that fell from heaven was as salty as salt water at sea." The storm, which raged from 9 A.M. until 10 P.M., washed away Fort Greene and its occupants on Cockspur Island; blew down the Presbyterian Church steeple; destroyed the Savannah exchange, jail, and courthouse; and flooded Hutchinson Island. From Savannah this storm continued northward, and its effects were felt as far north as Boston.

The next major hurricane to strike Georgia arrived in September 1824. This time, damage was especially high in Darien and the surrounding countryside. The *Darien Gazette* reported that two schooners were blown ashore, the eastern sawmill was in ruins, the rice fields were again inundated, the cotton crop was ruined, and fewer than six Darien houses had survived without damage. On St. Simons Island, one plantation reported "incalculable" damage "as the sea broke in and deluged the whole point, sweeping away buildings, undoing labor of years. . . ." On Sapelo Island, "a wall of water six feet high was swept across much of the island." This hurricane was reported as calamitous to the whole Georgia coast.

The violent hurricane of 1854 was a major storm that ravaged the entire east coast. As if Savannah did not have enough trouble in battling a raging yellow fever epidemic, this storm heaped on more. Trees were uprooted; Hutchinson Island was flooded once again; boats crashed into wharves, destroying both boats and wharves; a dry dock broke loose and floated downriver, wrecking ships along its path; and several people drowned.

The August 1881 hurricane, although barely a hurricane in terms of wind speed, was one of Georgia's most destructive. Savannah and Tybee again bore the storm's brunt, for it came ashore slightly north of Savannah, while the Brunswick area escaped relatively unscathed. The accompanying extremely high tides (12 feet above normal) caused much of the damage. At the coast more than 100 vessels were wrecked. In the city almost every building was damaged to some degree, and widespread debris completely blocked streets all over Savannah. Total damage in the city was assessed at $1.5 million. Worst of all, more than 330 people drowned in the extreme floodwaters.

On the ominous twelfth anniversary of the 1881 storm, the Savannah area was hit by yet another destructive hurricane. Like its predecessor, this August 1893 hurricane brought relatively mild winds but was accompanied by a devastating storm surge, this time 17 to 20 feet high. Between 1,000 and 2,000 people lost their lives as the tremendous surge of water submerged the Georgia and South Carolina islands and flooded low-lying mainland areas. Not surprisingly, the seawall at St. Simons Island was reported damaged. On Tybee Island, which was completely inundated, nearly every building was damaged, and the railroad to the island was wrecked. The other islands between the mainland and the ocean—Little Tybee, Wilmington, Talahi, Whitmarsh, Oatland, and Skidaway—also were flooded. $10 million in damage was recorded throughout the region.

Just three years later, the hurricane of September 1896 struck. Said to have wrought desolation beyond description in southeastern Georgia, this storm also brought heavy damages to the Savannah/Tybee area.

The year 1898 saw a number of hurricanes. The first storm struck late in August, wrecking harbors, washing out roads and railways, unroofing buildings, and downing wires throughout the Savannah/Tybee area. Barely a month later, the second storm of the season struck, bringing another disaster. The fury of this October hurricane made the previous two hurricanes seem almost mild. Hutchinson Island once again was flooded. Savannah got off relatively easy, sustaining damage that was widespread but less severe than in the previous storms. Brunswick and the southern islands, however, which were buffeted for a seemingly endless 18 hours, were not so fortunate. In its worst flood since 1812, Brunswick was inundated up to eight feet deep (fig. 3.23). What little of the rice crop had survived the August hurricane was totally destroyed. At Sapelo Island one resident of the southwest area reported that "waves coming across the island—direct from the ocean, covered the tops of our windows. In the house the water was nearly 3 feet deep—in the yard, nearly 6 feet deep on a level. The waves in our yard were fully 12 feet high." At least 120 deaths were reported in the Brunswick/Darien area, most of them by drowning at Butler Island. Some 80 more deaths were reported at the coastal islands near the mouth of the Altamaha River.

With the turn of the century, Georgia's luck with hurricanes took a turn for the better. The storms that struck were relatively infrequent and inflicted only mild damage. The August 1911 hurricane heavily damaged property on Tybee Island, and the 1928 hurricane destroyed beach cottages, downed many trees and utility lines, and damaged crops in the fields, but these damages were relatively inconsequential compared to the death and destruction wrought by the storms of the late 19th century.

The next storm to bring substantial damage came in October of 1947. With 100-mph winds, 15-foot waves, and a 12-foot storm surge, this hurricane substantially damaged more than 1500 buildings in Savannah Beach and Savannah.

Hurricane Dora (Category 2), which struck in September 1964, brought 90-mph winds and extremely high tides, causing heavy damage in coastal areas. Some buildings were extensively damaged, and island shorelines receded far enough that much beach property was undermined and washed away.

At least as significant as Dora's wind and water damage was the human response in its wake. Much of the riprap along the Georgia shore was placed there in reaction to the hurricane and is responsible today for the absence of a high-tide beach in many places (fig. 5.40).

The most recent hurricane to directly strike the Georgia coast was Hurricane David (Category 2) in September 1979. This storm made landfall in the vicinity of Ossabaw and St. Catherines Islands. It was a minor hurricane compared to storms of the late 19th century. Maximum winds were 90 mph offshore, 75 mph at Savannah, and 50 mph at Brunswick. Storm surge was only about 3.4 feet, in contrast to the 17- to 18-foot tide levels of some earlier storms. The wind-generated waves topped seawalls and revetments on the coastal islands and caused as much as 20 feet of dune recession on Jekyll. Even on Tybee, at the far northern end of the state, damage was estimated at $75,000 (even with the protection of a $3 million beach pumped up only three years earlier). Hurricane David may have changed offshore bottom configurations that allowed later northeasters to severely damage the coast.

Today, advance warning, efficient evacuation, and safer construction practices *should* result in low casualty rates even in a major hurricane. But unsafe development, allowing population growth to exceed the capacity for safe evacuation, and complacency on the part of coastal residents could reverse this trend with shocking results. Much can be learned from the devastation of Hurricane Hugo in South Carolina. The question is not whether such a storm can strike the Georgia coast, but when?

Appendix C   A Guide to Federal, State, and Local Agencies and Organizations Involved in Coastal Development

Numerous agencies at all levels of government are engaged in planning, regulating, or studying coastal development in Georgia. These agencies issue permits for various phases of construction and provide information on development to the homeowner, developer, or planner. Following is an alphabetical list of topics related to coastal development; under each topic are the names of agencies and organizations to consult.

Aerial Photography and Remote-Sensing Imagery

For a historical listing of available photography (type, scale, year flown, coverage, percentage of cloud cover, etc.), contact:

U.S. Department of Agriculture
Agricultural Stabilization and Conservation Service
Aerial Photography Field Office
P.O. Box 30010
Salt Lake City, UT 84130
(801) 524-5856

Local sources that may have aerial photographs of your area available for inspection include the office of the tax assessor in your county, libraries or departments of geology or geography in local colleges and universities, the Savannah District office of the U.S. Army Corps of Engineers, the Coastal Georgia Regional Development Center, and the Coastal Resources Division of the Georgia Department of Natural Resources (see under *Beach erosion*).

For information on satellite imagery contact:

EROS Data Center
Sioux Falls, SD 57198
(605) 594-6151

Archives and Records

Historical information on coastal areas may be obtained from offices of municipal and county clerks. Other sources include university libraries, local museums, local presses, and:

Tybee Island Museum
Box 366
Tybee Island, GA 31328
(912) 786-4077

Georgia Historical Society
501 Whitaker Street
Savannah, GA 31499
(912) 651-2128

Coastal Georgia Historical Society
Museum of Coastal History
P.O. Box 1136
St. Simons Island, GA 31522
(912) 638-4666

Federal Archives and Records Center
1557 St. Joseph Avenue
East Point, GA 30344
(404) 763-7474

Beach Erosion

Information on beach erosion, inlet migration, floods, and high winds is available from the district office of the U.S. Army Corps of Engineers. Contact:

U.S. Army Corps of Engineers, Savannah
P.O. Box 889
Savannah, GA 31402-0889
(912) 944-5822

For information on federal projects in the Savannah District, contact the Planning Division, Navigation and Coastal Section (912) 944-5378.

Two state sources exist for geological information. For publications, surveys, and general statewide information, contact:

Georgia Geologic Survey
Department of Natural Resources
Room 400
19 Martin Luther King Jr. Drive, S.W.
Atlanta, GA 30334
(404) 656-3214

For specific questions about beach erosion, rates, permits, and such, contact:

Marsh and Beach Section
Coastal Resources Division
Georgia Department of Natural Resources
1 Conservation Way
Brunswick, GA 31523-8600
(912) 264-7218

In early 1991 regional growth management planning was a state function. For information about planning, contact:

Coastal Georgia Regional
Development Center
Planning Division
P.O. Box 1917
Brunswick, GA 31521
(912) 264-7363

In addition, planning functions are handled by county and municipal agencies (such as the Chatham-Savannah Metropolitan Planning Commission).

Bridges and Causeways

The U.S. Coast Guard has jurisdiction over issuing permits to build bridges or causeways that will affect navigable waters. Information is available from:

Commander, Coast Guard District 7
909 S.E. 1st Avenue
Brickell Plaza Building
Miami, FL 33131-3050
(305) 536-5611

In the event of flooding and detours, contact your local police department.

Building Codes and Zoning

Georgia does not have a state building code. Contact your municipality or county for local codes and ordinances (Chatham County, for example, uses the Southern Standard Building Code, 1985 edition.)

Those communities participating in the National Flood Insurance Program will have building elevation requirements to meet the specifications of the program. For the specific code in your area, contact the city or county administrator.

Check to be sure that the property in which you're interested is zoned for your intended use and that adjacent property zones do not conflict with your plans. For information, contact the city or county building inspector.

Civil Preparedness

The Georgia Emergency Management Division handles civil preparedness at the state level; local Civil Defense organizations coordinate civil preparedness and emergency response activities. For more information, contact:

Georgia Emergency Management Agency
P.O. Box 18055
Atlanta, GA 30316-0055
(404) 624-6000

Chatham Emergency Management Agency
124 Bull Street
Suite 140
Savannah, GA 31401
(912) 651-3100

Brunswick/Glynn County Emergency Management Agency
P.O. Box 1021
Brunswick, GA 31521
(912) 267-5678

See also *Disaster Assistance*.

Coastal Zone Management

Georgia does not participate in the Federal Coastal Zone Management Program. The state's position has been that it can do better without federal interference. The Coastal Resources Division of the Georgia Department of Natural Resources is the lead agency in coastal natural resource management. For general information, contact their Marsh and Beach section. The address is listed under *Beach Erosion*.

The Coastal Georgia Regional Development Center is the lead agency in planning and guiding development and resource use. Their complete address is listed under *Beach Erosion*.

Although the state does not participate in the federal Coastal Zone Management Program, planners and managers may be interested in publications from the national Office of Ocean and Coastal Resource Management:

Office of Ocean and Coastal Resource
Management
National Oceanic and Atmospheric
Administration
1825 Connecticut Avenue, N.W.
Washington, DC 20236
(202) 606-4234

For information about the management of a particular island, contact:

Tybee Island:
City of Tybee Island
P.O. Box 128
Tybee Island, GA 31328
(912) 786-4573

Wassaw Island:
Wassaw Island National Wildlife Refuge
U.S. Fish and Wildlife Service
Savannah Coastal Refuge Complex
P.O. Box 8487
Savannah, GA 31412
(912) 944-4415

Williamson Island:
Office of the Commissioner
Georgia Department of Natural Resources
205 Butler Street
East Tower, Suite 1252
Atlanta, GA 30334
(404) 656-3500
or
Coastal Resources Division
Georgia Department of Natural Resources
1 Conservation Way
Brunswick, GA 31523-8600
(912) 264-7218

Ossabaw Island:
Ossabaw Island
Island Manager
Georgia Department of Natural Resources
Game Management Section
1 Conservation Way
Brunswick, GA 31523-8604
(912) 262-3174

St. Catherines Island:
St. Catherines Island Foundation
Route 1, Box 207-Z
Midway, GA 31320
(912) 884-5002

Blackbeard Island:
Blackbeard Island National Wildlife Refuge
U.S. Fish and Wildlife Service
Savannah Coastal Refuge Complex
P.O. Box 8487
Savannah, GA 31412
(912) 944-4415

Sapelo Island:
Sapelo Island Estuarine Research Reserve
Sapelo Island, GA 31328
or
Georgia Department of Natural Resources
Game Management Section
P.O. Box 15
Sapelo Island, GA 31327
(912) 485-2251

Little St. Simons Island:
Little St. Simons Island
P.O. Box 1079
St. Simons Island, GA 31522
(912) 638-7472

Sea Island:
Sea Island Company
100 First Street
Sea Island, GA 31561
(912) 638-3611

St. Simons Island:
Glynn County Planning Commission
Community Development Commission
P.O. 1495
Brunswick, GA 31523
(912) 267-5735

Jekyll Island:
Jekyll Island Convention and
Visitors Bureau
One Beachview Drive
P.O. Box 3186
Jekyll Island, GA 31520
(912) 635-2236

Little Cumberland Island:
Little Cumberland Island Association
P.O. Box 3127
Jekyll Island, GA 31520

Cumberland Island:
Cumberland Island National Seashore
P.O. Box 806
St. Marys, GA 31558
(912) 882-4335

For information about offshore environments, especially hard-bottom reef areas, contact:

Gray's Reef National Marine Sanctuary
National Oceanic and Atmospheric Administration
P.O. Box 13687
Savannah, GA 31416
(912) 598-2496

See also *Planning and Land Use* in this section.

Consultants

It is inappropriate for the authors of this book to recommend any individual or firm as a coastal or construction consultant, but we encourage prospective buyers as well as owners of existing property to seek expert advice on housing construction safety and site safety. The offices listed below under other topics and the offices of your local government are sources of advice on appropriate private consultants for your particular problem. *Low-Cost Shore Protection* by the U.S. Army Corps of Engineers (full reference given in appendix D, *Shoreline Engineering* section) contains brief sections on how to hire an architect, engineer, or contractor.

Disaster Assistance

In the event of a natural disaster, stay tuned to your local radio and television stations. Information and instructions regarding immediate disaster recovery will be broadcast from your local Civil Defense coordinator.

Emergency response functions such as flood warnings, evacuation, and damage assessments are handled at the state level by the Georgia Emergency Management Agency and are coordinated by local Civil Defense organizations (listed under *Civil Preparedness*). For information, contact:

Georgia Department of Defense
P.O. Box 17965
Atlanta, GA 30316
(404) 624-6000

The regional office of the Federal Emergency Management Agency can provide information about disaster relief, hazard mitigation, and insurance programs. For information, contact:

Federal Emergency Management Agency
Region IV
1371 Peachtree Street, N.E.
Suite 700
Atlanta, GA 30309
(404) 853-4200

For information about disaster assistance, contact Disaster Assistance Programs (404) 853-4302. For public buildings and communities, contact Public Assistance Branch (404) 853-4304; for individual homes, contact Individual Assistance Branch (404) 853-4308. For flood insurance questions, contact Natural and Technological Hazards Division (404) 853-4406.

The Natural Hazards Research and Applications Information Center serves as a national clearinghouse for information concerning social and public policy on natural hazards. It publishes *Natural Hazards Observer*, a quarterly newsletter, assists in organizing conferences, prepares bibliographies, and responds to specific information requests. For information, contact:

Natural Hazards Research and Applications Information Center
Campus Box 482
University of Colorado
Boulder, CO 80309-0482
(303) 492-6818

For warnings about major storms, see *Storm Warnings*.

Dredging, Filling, and Construction in Coastal Waterways

Georgia law requires that any person who wishes to remove, fill, dredge, drain, or otherwise alter any marshland within the estuarine area (tidally influenced) must obtain a permit from the Coastal Marshlands Protection Committee. Likewise, any shoreline engineering activity requires a permit under the Shore Assistance Act from the Shore Assistance Committee.

For more information, contact the Marsh and Beach section of the state Department of Natural Resources. The address is listed under *Beach Erosion*.

Federal law requires that any person who

wishes to dredge, fill, or place any structure in navigable water (almost any body of water) apply for a permit from the U.S. Army Corps of Engineers. Write for a useful information package on permits, but make sure you describe your proposed work in relation to wetlands and navigable waters. Information is available from the Regulatory Branch of the Corps' Operations Division at (912) 944-5348. The Savannah District address is listed under *Beach Erosion*.

Dune Alteration

Individual communities may have ordinances pertaining to dune alteration. Both Tybee Island and Glynn County have dune protection ordinances that are reinforced by the Shoreline Assistance Act. For information, call or write the local courthouse.

Emergency Management

See *Civil Preparedness* and *Disaster Assistance*.

Environmental Conservation and Education

For information about the marine science education programs of the University of Georgia and Sea Grant, contact:

University of Georgia Marine
Extension Service

P.O. Box 13687
Savannah, GA 31406
(912) 598-2496

Georgia Sea Grant College Program
University of Georgia
Ecology Building
Athens, GA 30602
(404) 542-7671

A public saltwater aquarium hall and a marine education field station with labs, classes, dormitory, and cafeteria are on campus at the University of Georgia's Marine Extension Service facility on Skidaway Island.

The Cooperative Extension Service of the University of Georgia operates marine science education centers at Tybee and Jekyll Islands. These centers offer overnight accommodations for groups, usually elementary school students. For more information, contact:

Camp Chatham 4-H Center on Tybee
9 Lewis Avenue
Tybee Island, GA 31328
(912) 786-5534

Jekyll Island 4-H Center
201 South Beachview Drive
Jekyll Island, GA 31527
(912) 635-2708

The Coastal Office of the Georgia Conservancy runs environmental education programs and offers publications about the Georgia coast. For more information, contact:

The Georgia Conservancy
711 Sandtown Road
Savannah, GA 31410
(912) 897-6462

The Georgia Environmental Council is a statewide coalition of organizations interested in environmental legislation. For more information, contact:

Georgia Environmental Council
3110 Maple Drive
Suite 407
Atlanta, GA 30305
(404) 993-7124

The Chatham County Board of Education opened the Oatland Island Education Center in 1974, and this unique educational facility now serves both school groups and the general public. Price of admission is a can of dog food, cat food, or bird-seed (preferably Alpo dog food or cat food). For more information, contact:

Oatland Island Educational Center
711 Sandtown Road
Savannah, GA 31410
(912) 897-3773

Geologic Information

For information:

U.S. Geological Survey
MS-119

Attn: Public Affairs
12201 Sunrise Valley Road
Reston, VA 22092
(703) 648-4460

Department of Natural Resources
Geologic Survey Branch
270 Washington St., S.W.
Atlanta, GA 30334
(404) 656-3214

See also *Beach Erosion*.

Groundwater

For information on groundwater resources in your area, contact:

Water Resources Management Branch
Environmental Protection Division
205 Butler Street, S.E.
Floyd Towers East
Atlanta, GA 30334
(404) 656-4807
or
District Chief
U.S. Geological Survey
Water Resources Division
6481 Peachtree Industrial Boulevard
Suite B
Doraville, GA 30360
(404) 986-6860

Hazards

See *Beach Erosion* and *Scientific Studies*. For hurricane warnings, see *Storm Warnings*.

History

See *Archives and Records*.

Housing and Insurance

Prospective homeowners should obtain some of the many references listed in appendix D under *Site Analysis* and *Building or Improving a Home*. When acquiring property, consider the following: (1) Descriptions and surveys of land in coastal areas are very complicated. Old titles granting fee-simple rights to property below the high-tide line may not be upheld in court; titles should be reviewed by a competent attorney before they are transferred. A boundary described as the high-water mark may be impossible to determine. (2) Ask about the provision of sewage disposal and utilities including water, electricity, gas, and telephone. (3) Be sure that any promises of future improvements (access, utilities, additions) are allowable under existing federal, state, and local regulations; coastal property is subject to more severe restrictions than inland property. (4) Be sure to visit the property and inspect it carefully (preferably during the stormy winter season) before buying it.

Subdivisions containing more than 100 lots and offered in interstate commerce must be registered with the Office of Interstate Land Sales Registration (as specified by the Interstate Land Sales Full Disclosure Act). Prospective buyers must be provided with a property report. This office also produces a booklet entitled *Get the Facts . . . Before Buying Land* for people who wish to invest in land. Information on subdivision property and land investment is available from:

Office of Interstate Land Sales Registration
U.S. Department of Housing and
Urban Development
451 7th Street, S.W., Room 6278
Washington, DC 20410
(202) 708-0502

For general information about the National Flood Insurance Program (NFIP) or inquiries about the laws, regulations, or administrative policies related to the NFIP, write:

Federal Emergency Management Agency
Federal Insurance Administration
500 C Street, S.W.
Washington, DC 20472

In coastal areas special building requirements must often be met to obtain affordable flood or windstorm insurance. To find out the requirements for your area, check with your insurance agent or obtain a copy of the local ordinance. Note that a flood policy under the national flood insurance program is separate from your regular homeowner's policy. For insurance questions, call local property insurance agents or brokers, or the National Flood Insurance Program offices, toll-free, 1 (800) 638-6620.

Your insurance agent or community building inspector should be able to provide you with information about the location of your building site on the Flood Insurance Rate Map (FIRM) and the elevation required for the first floor to be above the 100-year flood level.

To order Flood Hazard Boundary Maps, Flood Insurance Rate Maps, and for information on Flood Insurance Studies, call the Flood Map Distribution Center's toll-free number, 1 (800) 333-1363, or mail a Flood Insurance Map Order Form (obtained by calling the same number) to:

Federal Emergency Management Agency
Flood Map Distribution Center
6930 San Tomas Road
Baltimore, MD 21227

For information pertaining to hazard identification mapping and floodplain management, contact the FEMA/FIA address shown above or your FEMA Regional Office:

Federal Emergency Management Agency
Region IV
1371 Peachtree Street, N.E.
Suite 700
Atlanta, GA 30309
(404) 853-4400

For technical data that are the basis for flood hazard identification, write:

EDSP Repository
7500 Greenway Center Drive
Suite 700
Greenbelt, MD 20770

Hurricanes

The National Oceanic and Atmospheric Administration (NOAA) is a good agency from which to request information on hurricanes. To find out about available NOAA publications, contact:

Distribution Branch N/CG33
National Ocean Service, National Oceanic and Atmospheric Administration
Riverdale, MD 20737-1199
(301) 436-6990

See also *Storm Warnings* and *Movies and Audiovisual Materials*.

Insurance

See *Housing and Insurance*.

Land acquisition

See *Housing and Insurance*.

Maps

Maps are available to planners and managers and may be of interest to individual property owners. Topographic, geologic, land-use maps, and photo map quadrangles are available from:

U.S. Geological Survey
Earth Science Information Center
507 National Center
Reston, VA 22092
1 (800) USA-MAPS

A free index to the type of map desired (for example, the "Index to Topographic Maps of Georgia") should be requested and then used for ordering specific maps. Similar maps are available from the state geologist (see under *Geologic Information*).

Flood-zone maps: National Flood Insurance Program flood hazard boundary maps (FHBM) or flood insurance rate maps (FIRM) are generally on file with your local building official or can be obtained from FEMA's Flood Map Distribution Center, 1 (800) 333-1363. The address is listed under *Housing and Insurance*.

Planning maps: Call or write your local county planner (see under *Planning and Land Use*).

Georgia DNR also has available Resource Assessment Maps, Coastal Resource Maps, and Flood-Prone Area Maps. Contact:

Planning, Research and Evaluation
Georgia Department of Natural Resources
205 Butler Street, Suite 1252
Atlanta, GA 30334
(404) 656-6374

Nautical charts in several scales (1:20,000, 1:40,000, and 1:80,000) contain navigation information on Georgia coastal waters. These charts are also useful in documenting shoreline changes. A nautical chart index map is available from:

Distribution Branch N/CG33
National Ocean Service, NOAA
Riverdale, MD 20737-1199
(301) 436-6990

Charts of local areas may be obtained from local marinas. County and city street maps are available from local bookstores.

Marine and Coastal Zone Information

See under *Beach Erosion*, *Hazards*, and *Scientific Studies*.

Movies and Audiovisual Materials

Many of the audiovisual materials listed below may also be available for loan at coastal Georgia libraries.

The video *The Georgia Coast: Land, Sea, and Marsh* (1988) offers a general overview of the natural systems and processes of the Georgia coast. Produced by and available for purchase from:

University of Georgia Marine
Extension Service and Georgia
Sea Grant College Program
University of Georgia
Athens, GA 30602
(404) 542-7671

Guale is a documentary film (approximately one hour) about the history, people, and future of the Georgia coast. It is available for sale or rental from:

Southern Chroniclers
P.O. Box 8892
Savannah, GA 31401

The Georgia Department of Natural Resources has produced several award-winning films about important Georgia conservation issues. Of special interest to coastal residents: *The Spirit of Sapelo*, *Canada Geese: Georgia's Latest Arrival*, *Georgia: A Time for Choice* (about Georgia's endangered species), *Creatures of the Night: Georgia's Giant Sea Turtles*, and *A World Apart* (about coastal fishing). These films are available on loan, free of charge or for purchase ($20), from:

DNR Films
Georgia Department of Natural Resources
205 Butler Street, S.E., Suite 1258
Atlanta, GA 30334
(404) 656-0779

Other slide-tape programs and films on coastal topics are available from the National Audubon Society. *Coastal Follies* is a slide presentation that examines the problems of coastal development in light of the natural dynamics involved and offers guidelines for the benefit of financiers, developers, realtors, and current and prospective coastal property owners. Contact:

National Audubon Society
Southeast Regional Office
928 N. Monroe Street

Tallahassee, FL 32303
(904) 222-2473

The Beaches Are Moving (1990) looks at coastal issues through the eyes of Orrin Pilkey, Jr., coeditor of the Living with the Shore book series and "America's foremost philosopher of the beaches" (*New York Times Magazine*). Produced by the University of North Carolina Center for Public Television. Available for purchase from:

Environmental Media
P.O. Box 1616
Chapel Hill, NC 27514
(919) 933-3003

Sea Coast Crisis (1986) documents the degradation of our coastal waters as seen through the eyes of the fishermen and scientists who work in them. Available for purchase on videotape from:

Survival of the Sea Society
3299 K Street, N.W.
Sixth Floor
Washington, DC 20007

The movie *Portrait of a Coast* (1980) examines the effects of the rising sea level, wind, waves, and tides; it is available for purchase or rental from:

Jim Gabriel Productions
P.O. Box 400
Wellfleet, MA 02667

The film *The Atlantic's Last Frontier* examines Virginia's barrier island coast. Information on the film is available from:

The Nature Conservancy
1815 North Lynn Street
Arlington, VA 22209
(703) 841-5300

A slide program entitled *Barrier Islands and Beaches*, which describes many aspects of barrier islands from Massachusetts to Texas, is available from:

Dinesh C. Sharma
Environmental Consultant
2750 Rhode Island Avenue
Fort Myers, FL 33901
(813) 337-7199

The *Hurricane Preparedness Series* is a six-part video program about what to do in the event of a hurricane—both before and after. Each segment is 7–10 minutes long. Topics nclude (1) creating a hurricane preparedness plan, (2) evacuation, (3) planning to stay, (4) taking care of your boat, (5) taking care of your animals, and (6) the aftermath. For more information, contact the Public Information Director of the South Carolina Coastal Council:

South Carolina Coastal Council
4130 Faber Place   Suite 300
Charleston, SC 29405
(803) 744-5838

Shifting Sands is an excellent 28-minute videotape about Florida's coastal geologic history and the attempts to stabilize dynamic shores, but it also has great application to Georgia coastal issues. Designed for classroom use, the film is both entertaining and informative. For more information, contact:

Director
Florida Institute of Oceanography
830 First Street South
MSL 128B
St. Petersburg, FL 33701
(813) 893-9100

Parks and Recreation

For general tourist information, contact:

Georgia Tourist Division
Department of Community Development
270 Washington Street, S.W.
Atlanta, GA 30334
(404) 656-3590

For information about Georgia's coastal state parks or historic sites, contact:

Parks and Historic Sites Division
Georgia Department of Natural Resources
205 Butler Street
Suite 1352
Atlanta, GA 30334
(404) 656-2770
or

Coastal Resources Division
Georgia Department of Natural Resources
1 Conservation Way
Brunswick, GA 31523
(912) 264-7218

For information about Georgia forests, contact:

Georgia Forestry Commission
200 Piedmont Avenue
West Tower, Room 810
Atlanta, GA 30334
(404) 656-3204

For information about hunting and fishing in coastal Georgia, contact:

Game and Fish Division
Georgia Department of Natural Resources
205 Butler Street
Suite 1362
Atlanta, GA 30334
(404) 656-3523

Contact local fishing and hunting stores for additional information about available resources, clubs, and newsletters.

For information about Skidaway Island State Park:
Skidaway Island State Park
Savannah, GA 31411
(912) 598-2300

Jekyll Island, which is owned by the state and managed by the Jekyll Island Authority, is developed with many tourist amenities including motels, golf courses, a convention center, a fishing pier, a campground, picnic sites, bathhouses, and bicycle trails. For more information:

Jekyll Island Convention and Visitor Bureau
P.O. Box 3186
Jekyll Island, GA 31527
(800) 841-6586

For highway maps and information about road conditions:

Georgia Department of Transportation
2 Capitol Square
Atlanta, GA 30334
(404) 656-5267

For information about federal recreation areas, contact:

U.S. Forest Service Information Center
1720 Peachtree Road, N.W.
Atlanta, GA 30367
(404) 347-2384

For information about National Wildlife Refuges (on Blackbeard, Wassaw, and Wolf Islands) or other U.S. Fish and Wildlife Service holdings, contact:

U.S. Fish and Wildlife Service
Refuges and Wildlife, Room 1240
75 Spring Street, S.W.
Atlanta, GA 30303
(404) 331-0295
or

Savannah Coastal Refuges
P.O. Box 8487
Savannah, GA 31412
(912) 944-4415

For information about Cumberland Island National Seashore, contact:

National Park Service, Southeastern Office
75 Spring Street, S.W.
Atlanta, GA 30303
(404) 331-4998
or
Cumberland Island National Seashore
National Park Service
P.O. Box 806
St. Marys, GA 31558
(912) 882-4335

For information on diving off the Georgia coast, contact local dive shops. To obtain details on clubs, dive classes, dive trips, and dive sites, these stores offer perhaps the best information sources. The Gray's Reef National Marine Sanctuary Program can provide detailed information on the natural history and management of these areas. The complete address is listed under *Coastal Zone Management*.

Planning and Land Use

The Coastal Georgia Regional Development Center is responsible for coordinating the design and implementation of growth management strategy for the coastal region. Individual counties will guide development of the coastal region plan. Similar plans are to be developed throughout Georgia by other regional development centers. For information about this plan, contact the Planning Division of the Coastal Georgia Regional Development Center, at (912) 264-7363. The address is listed under *Beach Erosion*.

The Georgia Department of Community Affairs is responsible for coordinating work on the growth management strategy. Contact:

Georgia Department of Community Affairs
1200 Equitable Building
100 Peachtree Street
Atlanta, GA 30303
(404) 656-3836

For more information:

Governor's Office
203 State Capitol
Atlanta, GA 30334
(404) 656-1776

Local planning is also handled by the following:

Glynn County Planning Commission
P.O. Box 1495
Brunswick, GA 31521
(912) 267-5735

Savannah-Chatham County Metropolitan Planning Commission
P.O. Box 1027

Savannah, GA 31402
(912) 236-9523

See also *Beach Erosion* and *Coastal Zone Management*.

Roads and Property Access

Rules and regulations may vary from county to county and municipality to municipality. Before buying any property, determine what access rights and roads will be provided and what obligations you must meet.

For information about state highways, contact the Georgia Department of Transportation, listed under *Parks and Recreation*.

For local information, contact a county or municipal road official.

Tybee City limits:
Tybee Public Works-Sanitation Department
Polk Street
Tybee Island, GA 31328
(912) 786-5505

Chatham County (outside municipality limits):
Chatham County Public Works
P.O. Box 8161
Savannah, GA 31412
(912) 354-0402

Glynn County (outside of municipalities):
Glynn County Public Works Department
4145 Norwich Street

Brunswick, GA 31520
(912) 267-5760

Sanitation

Where property has no access to a sewer system, it is usually necessary to obtain a permit for a septic system from the local health department *before* a construction permit can be issued. Such a permit is issued only if the soil is suitable for a septic system. Clay-rich soils are usually unsuitable. Likewise, if your property does not have access to a municipal water system, you will need a well. Check with the county health department to determine the quality of the local groundwater. Make sure that the design and location of your septic system will safeguard your water supply.

Activity resulting in effluent discharge or runoff into surface waters requires certification from the state water pollution control agency that the proposed activity will not violate water quality standards. For information, contact:

Georgia Department of Public Health
878 Peachtree Street, N.E., Room 201
Atlanta, GA 30309
(404) 894-7505

A permit for the construction of a sewage disposal structure or any other structure in navigable waters of the United States must be obtained from the U.S. Army Corps of Engineers.

A permit for any discharge into navigable waters of the United States must be obtained from the U.S. Environmental Protection Agency. Recent judicial interpretation of the Federal Water Pollution Control Amendments of 1972 extends federal jurisdiction for protection of wetlands above the mean high-water mark. Federal permits may now be required for the development of land that is occasionally flooded by water draining indirectly into a navigable waterway. Information may be obtained from:

Water Protection Branch
Environmental Protection Division
205 Butler Street, S.E.
East Towers, Suite 1058
Atlanta, GA 30334
(404) 656-4708

Information is also available from:

Municipal Solid Waste Control Program
Environmental Protection Division
4244 International Parkway
Suite 100
Atlanta, GA 30354
(404) 362-2692

Check with your local county water and sewer department and municipal public services department to obtain more detailed information.

Scientific Studies

The University of Georgia Marine Institute conducts research into the biology, ecology, and geology of salt marshes, barrier islands, and the nearshore zone of the Georgia coast. For more information, contact:

University of Georgia
Marine Institute
Sapelo Island, GA 31327
(912) 485-2221

Marine Sciences Program
Ecology Building
University of Georgia
Athens, GA 30602
(404) 542-7671

University of Georgia Marine
Extension Service
Education Services Division
and Public Aquarium
P.O. Box 13687
Savannah, GA 31406
(912) 598-2496

University of Georgia Marine
Extension Service
Advisory Services Division
1715 Bay Street
Brunswick, GA 31520
(912) 264-7268

University of Georgia Marine
Extension Service
Seafood Technology and
Marketing Division
156 Trinity Avenue, S.W. Suite 303
Atlanta, GA 30303
(404) 331-4638

University of Georgia Marine
Extension Service
Shellfish Research Lab
P.O. Box 13687
Savannah, GA 31416
(912) 598-2348

Scientists at the Skidaway Institute of Oceanography conduct research in the Georgia coastal zone and other marine environments. The Institute's technical library is open on weekdays for reference only.

Skidaway Institute of Oceanography
P.O. Box 13687
Savannah, GA 31416
(912) 598-2453

Soils

Soil type is important in terms of (1) the type of vegetation it can support, (2) the construction technique best-suited to it, (3) drainage characteristics, and (4) ability to accommodate septic systems. Information and reports are available from:

U.S. Department of Agriculture
Soil Conservation Service
State Soil Scientist
Federal Building
3rd Floor, Box 13
355 East Hancock Avenue
Athens, GA 30601
(404) 546-2278

Soil Conservation Service—Savannah Office
USDA
124 Bernard Street
Suite 101
Savannah, GA 31401
(912) 236-0761

Georgia Department of Agriculture
Capitol Square
Atlanta, GA 30334
(404) 656-3600

Georgia State Soil and Water
Conservation Commission
P.O. Box 8024
Athens, GA 30603
(404) 542-3065

Storm Warnings

The U.S. Coast Guard is a source of information on ocean waves. For information, contact the local station.

General weather or climatic information for Chatham County may be obtained from:

National Weather Service
P.O. Box 7207
Savannah, GA 31408
(912) 964-2187

General weather or climatic information for other coastal counties is available from:

National Climatic Data Center
Federal Building
Asheville, NC 28801
(704) 259-0682

Hurricane information is available from:

National Oceanic and Atmospheric Administration
Office of Ocean and Coastal Resource Management
1825 Connecticut Avenue, N.W.
Washington, DC 20235
(202) 606-4234

The National Hurricane Center in Coral Gables, Florida, also offers a special telephone information service for named storms. For the latest storm advisory recorded by a forecaster from the National Hurricane Center, call: (900) 410-NOAA. The charge is $.50 for the first minute, plus $.35 for each additional minute.

For current weather and marine information forecasts, listen to your local radio and television stations or call the National Weather Service (Savannah) for a recorded message: (912) 964-1700.

Better yet, purchase a radio with the NOAA weather band. If an emergency arises, public information bulletins will be broadcast via the media and public address systems. Stay alert to local radio and television stations. Find out ahead of time which stations participate in the Emergency Broadcast System.

Vegetation

For information on the use of grass and other plantings for stabilization, consult the publication list in appendix D under *Vegetation*. Additional information on vegetation may be obtained from the agencies identified under *Dune Alteration*.

Wildlife

A variety of wildlife conservation and environmental organizations have local chapters in coastal communities. These groups can provide more information about local wildlife, as can local environmental education facilities.

For more information, contact the following:

National Audubon Society
Coastal Georgia Chapter
P.O. Box 1726
St. Simons Island, GA 31522
(912) 635-2495

Georgia Wildlife Federation
1936 Iris Drive
Conyers, GA 30207-5046
(404) 929-3350

The Georgia Conservancy, Inc.
781 Marietta Street, N.W.
Suite B100
Atlanta, GA 30318
(404) 876-2900

Wildlife Society, Georgia Chapter
Route 3, Box 180
Forsyth, GA 31029
(912) 994-1438

Game and Fish Division
Department of Natural Resources
205 Butler Street, S.E., Suite 1362
Atlanta, GA 30334
(404) 656-3523

U.S. Fish and Wildlife Service
Game and Fish Division
75 Spring Street, S.W., 12th Floor
Atlanta, GA 30303
(404) 331-0295

See also *Environmental Conservation and Education*.

Appendix D   Useful References

For the most recent publications, visit your coastal bookstore or contact the agencies and groups listed in appendix C. In this appendix, we include some general books, but emphasize the more technical literature and classic texts. Publications are listed by subject and appear in the approximate order in which these topics are addressed in this book.

Many of the publications distributed by government agencies are free or inexpensive. We encourage readers to pursue their interests and take advantage of these resources.

History

Indian Occupation and Geologic History of the Georgia Coast: A 5,000 Year Summary, by C. B. Depratter and J. D. Howard, 1980. Appears in *Excursions in Southeastern Geology: The Archeology-Geology of the Georgia Coast*, edited by J. D. Howard and others. Guidebook 20 of the Georgia Geologic Survey, prepared for 1980 Annual Meeting of the Geological Society of America. Available from the Georgia Survey Branch, 270 Washington St., S.W., Atlanta, 30334.

Guale Indians of the Southeastern United States Coast, by Grant D. Jones, 1980. Published in *Excursions in Southeastern Geology: The Archeology-Geology of the Georgia Coast*, edited by J. D. Howard and others. Guidebook 20 of the Georgia Geologic Survey, prepared for 1980 Annual Meeting of the Geological Society of America. Published by Georgia Department of Natural Resources. Available from the Geologic Survey Branch, 270 Washington St., S.W., Atlanta, 30334.

The Most Delightful Golden Islands, by Sir Robert Montgomery and Colonel John Barnwell, 1717. To create interest in the proposed settlement of Margravate of Azilia, this embellished account was published in what may have been the first "pitch" by a developer on the Georgia Coast. Originally published in London. Reprinted in 1909 by Cherokee Publishing Company, Atlanta.

General Oglethorpe's Georgia, Volume 1, by Mills B. Lane, 1975. This is a history of the first 10 years of the Colony of Georgia, from 1733 to 1743 as seen by General Oglethorpe, the colony's founder. This compilation of letters is the first of two volumes of Oglethorpe's correspondence. Published by Beehive Press, Savannah.

Historic Tybee Island, by Margaret Godley, 1958. Major aspects of Tybee's history, including military use, hurricanes, and recreational use are featured in this collection of anecdotes. Published by Savannah Beach Chamber of Commerce, Savannah Beach (now Tybee Island Chamber of Commerce, 209 Butler Avenue, Tybee Island, 31328).

St. Simons—Enchanted Island, by Barbara Hull, 1980. An enjoyable, easy-to-read account of the history of St. Simons Island. Published by Cherokee Publishing Company, Atlanta.

Sapelo: A History, by Buddy Sullivan, 1988. Described as "Being a full and documented account of the Thomas Spalding, Howard Coffin and Richard J. Reynolds eras of island ownership; extracts from the post-bellum Sapelo journal of Archibald C. McKinley; the island's main house; the Sapelo Island lighthouse; and Blackbeard Island." Published and distributed by McIntosh County Chamber of Commerce, Darien, Georgia.

This Happy Isle—The Story of Sea Island and the Cloister, by Harold H. Martin, 1978. Although some early historical material is included about island use prior to 1900, this book is primarily the story of the Sea Island Company. Interesting information and many anecdotes. Published by Sea Island Company, Sea Island, 31561.

Georgia's Land of the Golden Isles, 3rd edition, by Burnett Vanstory, 1981. An island-by-island history of the Golden Isles (Ossabaw Island through Cumberland Island) written by a resident of St. Simons Island. Focuses primarily on the social aspects of life on these islands since the arrival of Europeans. Published by University of Georgia Press, Athens. Available in paperback at local bookstores.

Population Growth and Income Distribution of Georgia's Counties, by Everett S. Lee and Douglas C. Bachtel, May 1985. Published in *Issues Facing Georgia*, vol. 2, no. 3, by College of Agriculture, Cooperative Extension Service, University of Georgia, Rural Development Center, Tifton.

A Demographic Study of the Coastal Coun-

ties of Georgia, 1790–1980, by D. Jedlicka, 1981. Published as Technical Report Series No. 81-3 by Georgia Marine Science Center, University of Georgia, Skidaway Island.

Coastal Georgia's Cultural Resources, by C. M. Howett, 1975. Classifies Georgia's major historical and archeological sites of natural or cultural significance according to coastal zone management requirements. Includes an article (pp. 182–199) on *The Value and Vulnerability of Coastal Resources*. Published by Georgia Department of Natural Resources, Atlanta.

Recreation

A Guide to the Georgia Coast, by The Georgia Conservancy, 1984. Excellent guide to the Georgia coastal area and the fun to be had there. Includes a quick introduction to the natural history of the area, as well as an extensive listing of natural, recreation, and historic sites (with hours of operation, admission fees, etc.). Available from The Georgia Conservancy, 711 Sandtown Road, Savannah, GA 31410.

Recreation on the Georgia Coast—An Ecological Approach, by Charles Clement and James Richardson, 1971. Outstanding article addressing the need to balance development, recreation, preservation, and fiscal responsibility at the coast. A bit old, but still carries a timely statement. Summary paper of a larger study by the Conservation Foundation. Published in *Georgia Business*, vol. 30, no. 11. Available at larger libraries.

Beachcomber's Guide to the Golden Isles, by Bertrand H. Dunegan, 1988. A short but thorough introduction to the organisms and shells on the beaches of Georgia and southern South Carolina. Available in most bookstores.

Recreation in the Coastal Zone, 1975. A collection of papers presented at a symposium sponsored by the U.S. Department of the Interior, Bureau of Outdoor Recreation, Southeast Region. Outlines different views of recreation in the coastal zone and the approaches taken by some states to recreation-related problems. The symposium was cosponsored by the Office of Coastal Zone Management (now Office of Ocean and Coastal Resource Management). Available from that office, National Oceanic and Atmospheric Administration, 1825 Connecticut Avenue, N.W., Washington, DC 20235.

Coastal Recreation: A Handbook for Planners and Managers, by Robert Ditton and Mark Stephens, 1976. Intended to provide technical assistance to planners and managers on major issues in coastal recreation, but may be of interest to property owners regarding public needs that may conflict with their rights to use the land. Published by U.S. Department of Commerce, National Oceanic and Atmospheric Administration, Office of Ocean and Coastal Resource Management, 1825 Connecticut Avenue, N.W., Washington, DC 20235.

Atlantic and Gulf Coasts, by W. H. Amos and S. H. Amos, 1985. One in the series of Audubon Society nature guides, this book covers many topics: geology, shells, seashore animals, mammals, fishes, reptiles and amphibians, plants, insects and spiders, and birds. Published by Alfred A. Knopf. Available in most libraries and bookstores.

The Audubon Society Field Guide to North American Seashells, by Harold A. Rehder, 1981. This well-illustrated reference is an excellent handbook for the serious shell collector. Published by Alfred A. Knopf. Available in most bookstores.

Field Guide to the Birds of North America, by the National Geographic Society, 1983. The newest field guide to the birds of North America and possibly the best. Unlike some, it uses excellent paintings commissioned for the work instead of color photos, and illustrates many different plumages showing marked differences due to age, sex, and geographical variation. The one bird guide to carry if you can carry only one. National Geographic Society, Washington, DC 20036. Available at most bookstores.

A Field Guide to the Birds of Eastern and Central North America, 4th edition, by Roger Tory Peterson, 1980. Provides brief species descriptions and detailed range maps for birds of the eastern United States, including shorebirds, wading birds, and waterfowl common to the Middle Atlantic region. Illustrated with color drawings. Published by Houghton Mifflin, Boston.

Top Flight Waterfowl Field Guide, by J. Rutheren and W. Zimmerman, 1979. Provides color illustrations of North American waterfowl species by color-coded groupings. Illustrations allow comparison of similar males and females and include pictures of each species in flight. Published by Moebiur Publishing Company, Milwaukee.

The Audubon Society Field Guide to North American Fishes, Whales, and Dolphins, by H. Boschung, J. Williams, D. Gotshall, D. Caldwell, and M. Caldwell, 1983. Provides detailed species accounts for fishes and marine mammals of North America. Illustrated with color photographs. Published by Alfred A. Knopf, New York.

Field Guide to Saltwater Fishes of North America, by A. McClane, 1978. Detailed family and species accounts for saltwater fish common to North America. Illustrated with color drawings. Published by Holt, Rinehart and Winston, New York.

The Audubon Society Field Guide to North American Seashore Creatures, by N. Meinkoth, 1981. Detailed species accounts and an overview of taxonomy of major shore animals of North America, including those common to the U.S. East Coast. Illustrated with color photographs. Published by Alfred A. Knopf, New York.

Dipping and Picking: A Guide to Recreational Crabbing, by D. C. Smith, 1978. How to catch, clean, and cook crabs. Available from Sea Grant Advisory Program, South Carolina Marine Resource Center, P.O. Box 12559, Charleston, SC 29412.

Barrier Islands

Straight Barrier Coasts: Cape Romain to Florida Keys, by F. P. Shepard and H. R. Wanless, 1971. Provides a detailed physical description of selected island and beach areas. Those Georgia islands included are Tybee, Ossabaw, Sapelo, and Blackbeard. In *Our Changing Coastline*, pp. 132–161, by F. P. Shepard and H. R. Wanless. Published by McGraw-Hill, New York.

Patterns and Trends of Land Use and Land Cover on Atlantic and Gulf Coast Barrier Islands, by H. F. Lins, 1980. U. S. Geological Survey Professional Paper 1156. A detailed description and analysis of changes in land use and land cover between 1945–1955 and 1972–1975 on each barrier island from Maine through Texas. Available from U.S. Government Printing Office, Washington, DC 20402.

Handbook of Coastal Processes and Erosion, edited by Paul Komar, 1983. A collection of technical papers (305 pages) dealing with a variety of coastal processes and environments. Of particular interest are chapter 5: "Barrier Islands" and chapter 6: "Patterns and Prediction of Shoreline Change." The former outlines barrier island dynamics, and the latter provides a summary of shoreline changes for Atlantic Coast barrier islands. Published by CRC Press, Boca Raton, FL 33431.

Barrier Islands and Beaches, 1976. Proceedings of the May 1976 barrier islands workshop. A collection of technical papers prepared by scientists studying islands. Provides a readable overview of barrier islands. Comprehensive coverage—from aesthetics to flood insurance—by experts. Topics include island ecosystems, ecology, geology, politics, and planning. Good bibliographic source for those studying barrier islands. Available from the Publications Department, Conservation Foundation, 1717 Massachusetts Avenue, N.W., Washington, DC 20036.

Coastal Geomorphology, edited by D. R. Coates, 1973. Collection of technical papers, including "Barrier Islands: Natural and Controlled" and "Comparison of Ecological and Geomorphic Interaction between Altered and Unaltered Barrier Island Systems in North Carolina." Interesting reading for anyone willing to overlook the jargon of coastal scientists. Published by State University of New York, Binghamton, NY 13901. Available only in larger university libraries.

Terrigenous Clastic Depositional Environments, edited by Miles Hayes and Tom Kana,

1976. Although compiled for a professional field course, this text provides an excellent detailed treatment on various coastal sedimentary environments that makes good reading for the interested nonscientist. The numerous photographs and diagrams support the text description of depositional systems in rivers, dunes, deltas, tidal flats and inlets, salt marshes, barrier islands, and beaches. Most of the examples are from South Carolina, but some can be applied to Georgia. Technical Report No. 11-CRD (184 pp.) of Coastal Research Division, Department of Geology, University of South Carolina, Columbia, SC 29208. Probably easiest to obtain through a college or university library.

Barrier Island Handbook, 2nd edition, by Steve Leatherman, 1982. A nontechnical, easy-to-read paperback about barrier island dynamics and coastal hazards. Many of the examples are from the Maryland and New England coasts but are of interest to Georgians. Available from Coastal Publications, 5201 Burke Drive, Charlotte, NC 28508.

Barrier Islands from the Gulf of St. Lawrence to the Gulf of Mexico, edited by Steve Leatherman, 1979. This collection of technical papers presents geological research on barrier islands. Published by Academic Press and available through most college and university libraries.

Sea-Level Changes

Policy Implications of Greenhouse Warming: The Full Report, by the National Academies of Sciences and Engineering, 1992. A collection of four reports (Natural Science, Mitigation, Adaptation, and Synthesis) reviewing the policy implications of global warming. Approximately 700 pages. Published by the National Academy Press, 2101 Constitution Avenue, N.W., Washington, DC 20418. Phone 1-800-624-6242.

Responding to Changes in Sea Level, Engineering Implications, by the Committee on Engineering Implications of Changes in Relative Mean Sea Level of the Marine Board Commission on Engineering and Technical Systems, National Research Council, 1987. Of interest to community planners, officials, and legislators because of the implications of sea-level rise for all coastal development. No specific solutions are provided, but the text does conclude with relevant general recommendations. Published by the National Academy Press, 2101 Constitution Avenue, N.W., Washington, DC 20418. Phone 1-800-624-6242.

Sea Level Variations for the United States 1855–1980, by S. D. Hicks, H. A. Debaugh, Jr., and L. E. Hickman, Jr., 1983. This government technical publication discusses trends in sea level since the mid-1800s to the present, based on the tide guage records. Published by National Oceanic and Atmospheric Administration, National Ocean Service, Rockville, MD 20852.

Sea Surface Studies: A Global View, edited by R. J. N. DeVoy, 1987. Excellent overview of sea-level changes, past and future. Although most of the discussion is quite technical, Part 4: *The Coastline: Processes, Planning, and Management* is accessible to the serious reader. Part 4 includes "Coastal Processes: The Coastal Response to Sea-Level Variation," "Man's Response to Sea-level Change," and "The Greenhouse Effect, Rising Sea Level, and Society's Response." Published by Croon Helm, Ltd., of New York, 649 pp. Available in libraries of larger universities or colleges.

New Greenhouse Report Puts Down Dissenters by Richard A. Kerr, 1990. An excellent, concise summary of the Scientific Assessment of Climate Change issued by a working group of the Intergovernmental Panel on Climate Change. Its important conclusion: there's virtual unanimity among climate experts that a warming is on the way and that the consequences will be serious. A sidebar, *The Greenhouse Consensus*, outlines those anticipated consequences. Published in *Science*, vol. 249, pp. 481–482. Available in many university libraries.

Coastal Environments

Barrier Beaches of the East Coast, by P. J. Godfrey, 1976. A clearly-written paper describing beaches and associated barrier island environments and processes. Published in *Oceanus*, vol. 19, no. 5, pp. 27–40. This issue was devoted to estuaries.

The Encyclopedia of Beaches and Coastal Environments, edited by M. L. Schwartz, 1982. This book is a good source of information about the coastal zone, but it is an expensive text aimed at the more serious student of the coast. Published by Hutchinson Ross, Stroudsburg, PA. Available at most college and university libraries.

At the Sea's Edge: An Introduction to Coastal Oceanography for the Amateur Naturalist, by W. T. Fox, 1983. Excellent, nontechnical, and richly illustrated introduction to coastal processes, meteorology, environments, and ecology. Published by Prentice-Hall, Englewood Cliffs, NJ 07632.

Edge of the Sea, by Russell Sackett, 1983. This volume in the Time-Life Planet Earth series outlines the importance and sensitivity of beaches and barrier systems. Coastal processes, the buffer-zone effect, the significance of coastal breeding grounds, and human impact on these environments are outlined in a non-technical presentation. Available through local bookstores or from Time-Life Books, 541 North Fairbanks Court, Chicago, IL 60611.

The Beachwalker's Guide: The Seashore from Maine to Florida, by E. Ricciuti, 1982. Easy-to-read description of the origins of the Atlantic coast's beaches, dunes, and marshes. Includes drawings, photographs, and descriptions of the dominant floral and faunal groups and species found along the Atlantic coast. Published by Doubleday and Company, Garden City, NY 11530.

Coastal Sedimentary Environments, edited by R. A. Davis, 1985. Provides background knowledge of coastal environments for scientists, engineers, planners, and managers. Oriented toward students of geology, but has a good mix of technical information and general description that makes it useful to anyone seriously interested in coastal processes. Published by Springer-Verlag, New York.

The Value and Vulnerability of Coastal Resources: Background Papers for Discussion and Review, prepared by Office of Planning, Research, and Evaluation, Georgia Department of Natural Resources, May 1975. This publication includes several papers that describe various facets of Georgia's coastal resources, including beach environments, terrestrial island ecology, fresh water systems, groundwater resources, coastal wildlife, cultural resources, soils, vegetation, salt marshes, and open marine and estuarine waters. Available only in major libraries.

General Recommendations, by H. T. Odum and others, 1974. Defines needs associated with coastal ecological systems. Of special interest are the discussion of a zoned-sector plan for the multiple development of bays and a zoning example on the Georgia coast. Pp. 132–141 in *Coastal Ecological Systems of the United States, Volume 1*. Published by Conservation Foundation, 1717 Massachusetts Avenue, N.W., Washington, DC 20036.

An Ecological Survey of the Coastal Region of Georgia, by A. S. Johnson and others, 1974. Describes the ecological characteristics of Georgia's coastal region. Specific chapters focus on the fauna, vegetation, and productivity of the islands, marshes, and open marine and estuarine waters. Emphasis on the islands of Tybee, Wassaw, Ossabaw, St. Catherines, Blackbeard, Sapelo, Little St. Simons, St. Simons, Jekyll, and Cumberland. National Park Service's Scientific Monograph Series no. 3.

Terrestrial Ecology of the Georgia Barrier Islands, by J. I. Richardson and J. S. Worthington, 1975. Summarizes the terrestrial ecology and physical environment of the Georgia barriers and the survival strategies of the island inhabitants. Appears in *The Value and Vulnerability of Coastal Resources*, pp. 35–111. Published by the Georgia Department of Natural Resources, Atlanta, 30334. Available in libraries of larger colleges and universities.

Ecological Characterization of the Sea Island Coastal Region of South Carolina and Georgia Resource Atlas, edited by Jane Davis and others, 1980. This excellent collection of oversize maps, charts, diagrams, and photographs tells just about everything you want to know about the Sea Island region (ecology, physiography, geology, climatology, and cultural and natural resources). The volume is a compilation of many investigations prepared for the U.S. Environmental Protection Agency and the Fish and Wildlife Service, U.S. Department of the Interior. The report *FWS/OBS-

79/43 is produced through the Biological Services Program and Interagency Energy-Environment Research and Development Program. Given its large size and limited printing, persons wishing to examine a copy should try larger libraries or one of the coastal agencies listed in appendix C.

Existing Aerial Photographic Resources of Coastal Georgia and a Brief Listing of Interpretive Aids, by R. A. Lindhurst and R. Reimold, 1973. This 1973 summary of (then available) photographic data for Georgia's coast includes detailed information on acquisition, interpretation, and use of photographic materials. Photographic equipment, aeronautical charts, topographical maps, tidal data, and other resources are described. Technical advances have rendered some of its information obsolete, but this publication still provides an introduction to historical data. Georgia Marine Science Center Technical Report 73-4, Skidaway Island. Available at larger Georgia libraries.

Beaches

"Beach Erosion on the Georgia Coast," by J. D. Howard and others, 1980. An unpublished paper of Skidaway Institute of Oceanography, Savannah, 31416.

Waves and Beaches, revised and updated edition, by Willard Bascom, 1980. A discussion of beaches and coastal processes. Published by Anchor Books, Doubleday and Company, Garden City, NY 11530. Available in paperback from local bookstores.

Beaches and Coasts, 2nd edition, by C. A. M. King, 1972. Classic treatment of beach and coastal processes. Published by St. Martin's Press, 175 Fifth Avenue, New York, NY 10010.

Beach Processes and Sedimentation, by Paul Komar, 1976. Technical explanations of beaches and beach processes. Recommended to serious students of the beach. Published by Prentice-Hall, Englewood Cliffs, NJ 07632.

Atlantic Beaches, by J. N. Leonard, 1972. The aesthetics of the beach in words and pictures. Published as part of the American Wilderness Series by Time-Life Books, Rockefeller Center, New York, NY 10020.

Vegetation

A Smithsonian Guide to Seaside Plants of Gulf and Atlantic Coasts, by Wilbur Duncan and Marion Duncan, 1987. This pictorial key to plants of dune environments, interdune areas, and dune-forest transition zones serves as an excellent and easy-to-use source for identification of seaside plants. Covers the area from Louisiana to Massachusetts (with the exception of south Florida). Published by the Smithsonian Institution Press, 470 L'Enfant Plaza, Suite 7100, Washington, DC 20560.

Manual of the Vascular Flora of the Carolinas, by A. Radford, H. Ahles and C. R. Bell, 1968. This key to Carolina plants is applicable to the coast of Georgia. For the amateur or professional botanist, the book is an indispensable tool for identifying plant species. Published by University of North Carolina Press, Chapel Hill, NC 27514.

Seacoast Plants of the Carolinas for Conservation and Beautification, by K. E. Braetz, 1973. An excellent discussion of plants native to beach and dune environments. Suggestions for plantings to stabilize and protect dunes and to landscape the beach. Evaluations of perennial beach plants. Descriptions and illustrations of various natural and ornamental plants. Written for the Carolinas, but applicable to Georgia. Available from coastal offices of the District Conservationists or from UNC Sea Grant, North Carolina State University, Box 8605, Raleigh, NC 27695-8605.

The Dune Book: How to Plant Grasses for Dune Stabilization, by Johanna Seltz, 1976. Brochure outlining the importance of sand dunes and means of stabilizing them through grass plantings. Available from UNC Sea Grant, North Carolina State University, Box 8605, Raleigh, NC 27695-8605.

Shore Stabilization with Salt Marsh Vegetation, by P. L. Knutson and W. W. Woodhouse, Jr., 1983, and *Shore Erosion Control with Salt Marsh Vegetation,* by P. L. Knutson and M. P. Inskeep, 1982. Summarize the use of coastal marsh vegetation as an erosion control measure. Artificial plantings are often a good alternative to building protective structures against erosion of low-energy or sheltered shorelines, such as in bays, sounds, lagoons, and estuaries. These publications outline criteria for determining site suitability for planting, selection of plant types, planting procedures, and estimating costs. These reports and additional information on coastal stabilization are available from the U.S. Army Corps of Engineers, Coastal Engineering Research Center, P.O. Box 631, Vicksburg, MS 39180, or from National Technical Information Service (NTIS), Attn: Operations Division, 5285 Port Royal Road, Springfield, VA 22161. Request publication by title, author, and year.

Vegetation Establishment and Shoreline Stabilization: Galveston Bay, Texas, by J. W. Webb and J. D. Dodd, 1976. This study evaluates plants as shoreline stabilizers in a low-energy estuarine environment in Texas. The results may be pertinent to similar environments along the Georgia shore. Available as CERC Technical Paper 76-13 from the U.S. Army Corps of Engineers, Coastal Engineering Research Center, P.O. Box 631, Vicksburg, MS 39180, or from NTIS, Attn: Operations Division, 5285 Port Royal Road, Springfield, VA 22161.

Building and Stabilizing Coastal Dunes with Vegetation (UNC-SG-82-05) and *Planting Marsh Grasses for Erosion Control* (UNC-SG-81-09), by S. W. Broome, W. W. Woodhouse, Jr., and E. D. Seneca, 1982. These publications on using vegetation as stabilizers are available from UNC Sea Grant, North Carolina State University, Box 8605, Raleigh, NC 27695-8605. Include publication number with your request.

Dune Building and Stabilization with Vegetation, by W. W. Woodhouse, Jr., 1978. This report includes a section on plants and planting methods used to build and stabilize dunes. Available from the Superintendent of Documents, U.S. Government Printing Office, Washington, DC 20402. Request Stock No. 008-022-00124-7.

Artificial Seaweed for Shoreline Erosion Control?, by Spencer Rogers, Jr., 1986. An excellent summary of experience the world over with artificial seaweed. Concludes that artificial seaweed, in general, is not an effective means of erosion control. UNC Sea Grant Publication UNC-SG-WP-86-4, UNC Sea Grant, North Carolina State University, Box 8605, Raleigh, NC 27695-8605.

Storms

Georgia Tropical Cyclones and Their Effect on the State, by Horace S. Carter, 1970. An interesting historical survey by the Georgia State Climatologist. Published by Environmental Science Services Administration, Environmental Data Service, Wheaton, MD 20902.

Tropical Cyclones of the North Atlantic Ocean, 1871–1977, by C. J. Neumann, G. W. Cry, E. L. Caso, and B. R. Jarvinen, 1978. This report gives useful information about the characteristics and classification of tropical cyclones. Storm tracks are also included. Published by National Oceanic and Atmospheric Administration. 170 pp. Available from U.S. Government Printing Office, Washington, DC 20402. Refer to stock number 003-017-00425-2.

Early American Hurricanes, 1492–1870, by D. M. Ludlum, 1963. Informative and entertaining descriptions of hurricanes affecting the Atlantic and Gulf coasts. Storm accounts in chronological order provide insight into their frequency, intensity, and destructive potential. Published by American Meterological Society, Boston. Available in public and university libraries.

Atlantic Hurricanes, by G. E. Dunn and B. I. Miller, 1960. Discusses at length hurricanes and associated phenomena such as storm surge, wind, and sea action. Includes a detailed account of Hurricane Hazel, 1954, and suggestions for pre- and posthurricane procedures. Published by Louisiana State University Press, Baton Rouge, LA 70893. Available in public and college libraries.

The Hurricane and Its Impact, by R. H. Simpson and H. Riehl, 1981. Discusses hurricane origin; impact of wind, waves, and tides; assessment and risk reduction; awareness and preparedness; prediction and warning; plus informative appendixes. This volume should be in libraries of coastal communities. Published by Louisiana State University Press, Baton Rouge, LA 70893.

Atlantic Hurricane Frequencies along the

U.S. Coastline, by R. H. Simpson and M. B. Lawrence, 1971. National Oceanic and Atmospheric Administration Technical Memorandum NWS SR-58. Available in large university libraries.

Storm Tide Frequency Analysis for the Coast of Georgia, by F. P. Ho, 1974. Technical Memorandum NWS HYDRO-19 of the Office of Hydrology, National Oceanic and Atmospheric Administration, Silver Spring, MD 20910.

Hurricanes and Coastal Storms: Awareness, Evacuation, and Mitigation, edited by E. J. Baker, 1980. Collection of reports addressing the problems associated with coastal storms. The paper "Living with Coastal Storms: Seeking an Accommodation" pp. 4–10, is of particular interest. This 219-page collection is available as Report No. 33 of the Florida Sea Grant College from Marine Advisory Program, University of Florida, Gainesville, FL 32611.

Storms, People and Property in Coastal North Carolina, by Simon Baker, 1978. Although this publication is aimed at property owners on North Carolina's barrier islands, it will be of interest to homeowners on Georgia's open-ocean coast. Topics include storm frequency and preparations for storm impact. Available as Sea Grant Publication UNC-SG-78-15 from UNC Sea Grant, North Carolina State University, Box 8605, Raleigh, NC 27695-8605.

Forecasting Extratropical Storm Surges for the Northeast Coast of the United States, by N. A. Pore, W. S. Richardson, and H. P. Perrotti, 1974. This 70-page technical report describes northeasters and the associated storm surge in nearshore areas. Several individual storms are described, and the Ash Wednesday storm of March 1962 receives special attention. Published as NOAA Technical Memo NWS TDL-50 by the National Weather Service. Available through larger libraries.

The Strategic Role of Perigean Spring Tides in Nautical History and North American Coastal Flooding, 1635–1976, by F. J. Wood, 1978. This 538-page report reviews tides and tidal forces with emphasis on perigean tides. These maximum spring tides occur when the gravitational pull of the sun and the moon is at a maximum. Although the report is largely technical, it includes a nontechnical discussion and historical review, along with an interesting collection of newspaper clippings. The Ash Wednesday storm of March 1962 is one of the many examples provided. Prepared for the National Oceanic and Atmospheric Administration and available from the Superintendent of Documents, U.S. Government Printing Office, Washington, DC 20402. Refer to stock no. 003-017-00420-1.

The Ash Wednesday Storm, by David Stick, 1987. Excellent photographic and anecdotal documentation of the effects of the 1962 Ash Wednesday storm on the Outer Banks of North Carolina. This 100-page book clearly conveys the messages that (1) powerful ocean forces can bring sudden and dramatic changes to the islands, and (2) careful site selection *does* make a difference. Published by Gresham Publications, P.O. Drawer 804, Kill Devil Hills, NC 27948.

Coastal Storm Awareness and Preparedness

Hurricane Information and Atlantic Tracking Chart, by the National Oceanic and Atmospheric Administration. A brochure that describes hurricanes, defines terms, lists safety rules. Outlines the method of tracking hurricanes and provides a tracking map. Available from Superintendent of Documents, U.S. Government Printing Office, Washington, DC 20402.

Hurricane Survival Checklist. Free publication available from the Insurance Information Institute. Send a self-addressed, stamped, business-size envelope to Publications Service Center, Insurance Information Institute, 110 William Street, New York, NY 10038.

Safety Tips for Hurricanes, by the Federal Emergency Management Agency, 1980. Leaflet of hurricane safety tips, available as Publication No. L-105 from FEMA, Washington, DC 20472. Also available as Stock No. 1984 0-454-626 from Superintendent of Documents, U.S. Government Printing Office, Washington, DC 20402.

Application of the National Flood Insurance Program to a Coastal Environment, by the Coastal Area Planning and Development Commission (CAPDC), 1981. Available from Coastal

Georgia Regional Development Center, P.O. Box 1917, Brunswick, GA 31521.

Glynn County Hurricane Response Plan, by the Coastal Area Planning and Development Commission, 1981. Available from Coastal Georgia Regional Development Center, P.O. Box 1917, Brunswick, GA 31521.

Surviving the Storm, 1981. An excellent 12-page summary of storm precautions prepared for the Virgin Islands, but useful anywhere. Available from "Surviving the Storm," P.O. Box 1208, St. Thomas, U.S. Virgin Islands 00801.

Preparing for Hurricanes and Coastal Flooding: A Handbook for Local Officials, 1983. A comprehensive handbook, valuable for all phases of local, county, and state planning for coastal storms. Order from Superintendent of Documents, U.S. Government Printing Office, Washington, DC 20402 (publication no. 1985-0-419-938/28).

Shoreline Engineering

Beach Nourishment Along the Southeast Atlantic and Gulf Coasts, by Todd Walton and James Purpura, 1977. Article in *Shore and Beach* magazine, pp. 10–18 (July), examines successes and failures of several beach nourishment projects.

Beach Behavior in the Vicinity of Groins, by C. H. Everts, 1979. An interesting description of the effects of two groin fields in New Jersey, which concludes that groins deflect the movement of sand seaward, causing erosion in the downdrift shadow area. This negative downdrift effect occurs even if groin compartments are filled with sand. Published in the *Proceedings of the Speciality Conference on Coastal Structures '79* (pp. 853–867) and available from U.S. Army Coastal Engineering Research Center, Waterways Experiment Station, P.O. Box 631, Vicksburg, MS 32180-0631. Ask for Reprint 79-3.

Low-Cost Shore Protection, by the U.S. Army Corps of Engineers, 1981. A set of four reports written for the layman includes the introductory report, a property owner's guide, a guide for local government officials, and a guide for engineers and contractors. The reports are a summary of the Shoreline Erosion Control Demonstration Program and suggest a wide range of engineering devices and techniques, including beach replenishment and vegetation, to stabilize shorelines. In adopting these approaches, you should keep in mind that they are short-term measures and may have unwanted side effects. The reports are available from Section 54 Program, U.S. Army Corps of Engineers, USACE (DAEN-CWP-P), Washington, DC 20314.

Shoreline Waves: Another Energy Crisis, by Victor Goldsmith, 1975. Shelf bathymetry is shown to be a controlling factor in wave refraction, which, in turn, controls wave height distribution along the beach. Notes that wave energy distribution may be controlled by modifying bathymetry. Available free from Sea Grant College Program, Virginia Institute of Marine Science, Gloucester Point, VA 23062. Request VIMS Contribution No. 734.

Shore Protection Guidelines, by the U.S. Army Corps of Engineers, 1971. Summary of the effects of waves, tides, and winds on beaches and engineering structures used for stabilization. Available free from Department of the Army, Corps of Engineers, Washington, DC 20318.

Shore Protection Manual, by the U.S. Army Corps of Engineers, 1984. The "bible" of shoreline engineering. Published in two volumes. Request publication 008-022-00218-9 from Superintendent of Documents, U.S. Government Printing Office, Washington, DC 20402.

Help Yourself, by the U.S. Army Corps of Engineers, 1978. Brochure addressing the erosion problems in the Great Lakes region. Will be of interest to island dwellers as it outlines shoreline processes and illustrates a variety of shoreline engineering devices used to combat erosion. Available free from U.S. Army Corps of Engineers, North Central Division, 219 South Dearborn Street, Chicago, IL 60604.

Bibliography of Publications Prior to July 1983 of the Coastal Engineering Research Center and the Beach Erosion Board, by A. Szuwalski and S. Wagner, 1984. A list of published coastal research by the U.S. Army Corps of Engineers. Available free from Coastal Engineering Research Center, U.S. Army Engineer Waterways Experiment Station, P.O. Box 631, Vicksburg, MS 39180.

List of Publications of the U.S. Army Engi-

neer Waterways Experiment Station, Volumes I and II, by R. M. Peck, 1984 and 1985. Update of the above bibliography and list of publications by other research branches of the Waterways Experiment Station. Available free from Special Projects Branch, Technical Information Center, U.S. Army Engineer Waterways Experiment Station, P.O. Box 631, Vicksburg, MS 39180.

Construction Materials for Coastal Structures, by Moffatt and Nichol, Engineers, 1983. Lengthy (427 pp.) summary of the characteristics of a wide range of materials used in coastal structures, beach protection devices, and erosion control. This technical reference guide should particularly interest coastal engineers and construction contractors who build such structures. Request Special Report No. 10 from Coastal Engineering Research Center, U.S. Army Engineer Waterways Experiment Station, P.O. Box 631, Vicksburg, MS 39180.

Annual Report of the Chief of Engineers on Civil Works Activities, by the U.S. Army Corps of Engineers, published annually since 1850. Summarizes federal activities related to jetty construction, dredging, harbor improvements, beach replenishment, and other shore stabilization projects. Available for viewing at federal depositories and Corps of Engineers District offices. (The district office in Georgia is in Savannah.) Also available for sale from Superintendent of Documents, U.S. Government Printing Office, Washington, DC 20402.

House and Senate Documents and Reports, by the U.S. Army Corps of Engineers. If a Corps of Engineers report to Congress contains particularly useful information, Congress may elect to have the Government Printing Office publish it. These reports contain much valuable information not available anywhere else. A librarian at a federal depository can help you locate documents and reports on your particular area of interest.

Geology and Land Use

The Beaches Are Moving: The Drowning of America's Shoreline, by Wallace Kaufman and Orrin Pilkey, 1979. This highly readable account of the state of America's coastline explains natural processes at work at the beach, provides a historical perspective of man's relation to the shore, and offers practical advice on how to live in harmony with the coastal environment. Originally published by Anchor Books/Doubleday, it is available in paperback with a 1983 epilogue by the authors from Duke University Press, 6697 College Station, Durham, NC 27708.

Coastal Dunes: Their Function, Delineation, and Management, by Paul Gares, Karl Nordstrom, and Norbert Psuty, 1979. This technical report provides information about the basic factors that influence dune formation. It also provides a methodology for defining the boundaries of a management zone along the shoreline in which land use would be controlled, and it discusses the rationale for this nonengineering approach to coastal protection. Available from Center for Coastal and Environmental Studies, Rutgers University, New Brunswick, NJ 08903.

Barrier Beach Development: A Perspective on the Problem, by S. P. Leatherman, 1981. This article in *Shore and Beach* magazine (vol. 49, no. 2, pp. 2–9) gives an excellent account of the problems of building in the coastal zone. The brief, nontechnical discussion includes an outline of the federal government's role and offers recommendations.

Hazards

A Coordination, Education and Mitigation Model for Disaster Preparedness in Coastal Areas, by the Coastal Area Planning and Development Commission, 1980. Available from Coastal Georgia Regional Development Center, Brunswick, 31521.

Flood Hazard Management Profiles, 1984. Center for Urban and Regional Studies, University of North Carolina, Chapel Hill, NC 27514.

Individual Islands

Historical Changes in the Mean High Water Shoreline of Georgia, 1857–1982, by M. M. Griffin and V. J. Henry, 1984. An invaluable, well-illustrated summary of historical shoreline changes on the coast of Georgia. Chapter

5 of this book draws heavily from this bulletin. Bulletin 98 of Georgia Department of Natural Resources, Environmental Protection Division, Georgia Geologic Survey, Atlanta 30334.

The Status of the Barrier Islands of the Southeastern Coast, by Langdon Warner, 1976. A readable, useful reference. General information on island environments, development pressures, government stimulants to private development, and property assessments. Tables summarize the status of development on barrier islands in each southeastern coastal state. Those seeking more detailed information on the development status, property assessments, and local land-use regulations on individual islands will want to obtain *Barrier Island Inventory*. These references, however, will be somewhat dated, given the past decade's burst of coastal development. Both references available from Open Space Institute, 36 West 44th Street, Room 1018, New York, NY 10036.

Beach Erosion and Hurricane Study of the Shores of Georgia and Portions of Florida, by the Beach Erosion Board, 1961. An unpublished paper available at U.S. Army Corps of Engineer District Office, Savannah.

A Study of Beach Erosion and Beach Erosion Control Efforts on Tybee, St. Simons, and Jekyll Islands, Georgia, and Amelia Island, Florida, by Lewis A. Taylor, 1981. An unpublished survey prepared for James Howard, Skidaway Institute of Oceanography, Savannah. Available for review at Skidaway Institute of Oceanography library.

Tri-state Conference Report: Methods for Beach and Sand Dune Protection, March 31–April 2, 1974, Jekyll Island, Georgia. Conference proceedings (48 pp.) discuss the importance of the natural beach/dune system, typical features of a Georgia barrier island, the role of dunes in shoreline stability, and related issues. References made to Savannah Beach, Tybee Island. Published by Georgia Department of Natural Resources, Atlanta 30334.

Savannah Beach—Plan for a Comprehensive Future, 1975. The first land-use plan for Tybee Island. Available at Tybee Island public library and at City Hall.

A Comprehensive Plan for Tybee Island, Georgia, 1985–2005, by J. Dukes and others, 1985. A thorough, balanced study with specific land-use recommendations based on consideration of future growth, population changes, the economy, local tourism-based businesses, erosion control, parking and beach access, land-use plans and zoning ordinances, drainage, and public services. Produced by a group of volunteer planners and published by Municipal Planning Commission, Tybee Island. Available at the Tybee Island public library or at City Hall.

Recommendations for Long-Term Mitigation of Beach Erosion on Tybee Island, Georgia: Report of the Tybee Island Technical Task Force, by the Tybee Island Technical Task Force, 1984. A 19-page report reviewing the condition of the beach at Tybee Island and recommending (1) protection of the seawall, (2) construction of the south end groin (now completed), and (3) renourishment and monitoring of Tybee's oceanfront beach. Submitted to Commissioner of the Department of Natural Resources, 270 Washington Street, S.W., Atlanta, 30334. Available at Tybee Island library.

History of Erosion and Erosion Control Efforts at Tybee Island, Georgia, by G. F. Oertel and others, 1985. Complete review of federal efforts at erosion "control" on Tybee Island. The section listing historical construction projects usefully documents the "New Jerseyization" of Tybee Island. Published by U.S. Army Corps of Engineers, Waterways Experiment Station, Coastal Engineering Research Center, Vicksburg, MS. Available from National Technical Information Service, 5825 Port Royal Road, Springfield, VA 22151.

Case Study on Fifty Years of Seawall and Groin Construction, Tybee Island, Georgia, by F. H. Posey, Jr. and J. A. Dick, 1987. Brief paper that "discusses and summarizes 86 years of man's struggle against Nature in the hostile environment of the coastal zone along the North Atlantic coast of South Georgia," (from the author's abstract). Published in *Coastal Zone '87*, vol. 3, pp. 3221–3232, by American Society of Civil Engineers, New York, 10017.

Unpredicted Rapid Erosion, Tybee Island, Georgia, by F. H. Posey, Jr., and F. W. Seyle, Jr., 1980. Readable paper relating in detail the early history of the 1976 Tybee Island beach replenishment project. Published in *Coastal Zone '80*, vol. 3, pp. 1869–1882, by American Society of Civil Engineers, New York, 10017.

Patterns of Water Flow and Sediment Dispersion Adjacent to an Eroding Barrier Island, by G. F. Oertel, 1973. Technical Report Series 73-9 of Georgia Marine Science Center, University of Georgia, Savannah, 31406.

Sedimentary Framework of the Eroding Beach-Shoreface System, Adjacent to Tybee Island, Georgia, by G. F. Oertel, 1973. A technical paper (30 pp.) that presents basic data on the Savannah Beach sediment budget and detailed topographic surveys of the shoreface. Sea Grant Technical Report Series No. 73-5. Published by University of Georgia, Georgia Marine Science Center, Savannah, 31406.

Patterns of Sediment Dispersion on the Shorelines of an Eroding Barrier Island, by G. F. Oertel, 1974. A 90-page technical report that discusses sediment transport patterns at Tybee Island and their relation to erosion patterns there. Sea Grant Technical Report Series No. 74-2. Published by University of Georgia, Georgia Marine Science Center, Savannah, 31406.

Tybee Island, Georgia, Beach Erosion Control Project (Final Environmental Impact Statement), by the U.S. Army Corps of Engineers, Savannah District, 1973. Describes the beach replenishment and groin project at Tybee Island. Available at Corps of Engineers Savannah District Office, Savannah, 31402.

Special Flood Hazard Information for the City of Savannah Beach, Georgia, by the U.S. Army Corps of Engineers, Savannah District, 1972. Available at Corps of Engineers Savannah District Office, Savannah, 31402.

Tybee Island, Georgia, by the U.S. Army Corps of Engineers. A 1959 investigation of hurricane problems at Tybee Island. Available at Savannah District of U.S. Army Corps of Engineers or as House Document #354/86/2 at federal depository libraries.

Preliminary Tidal Flood Information for the Coastal Area of Chatham County, Georgia, by the U.S. Army Corps of Engineers, 1968. Published by the Savannah District of the U.S. Army Corps of Engineers and available there.

Differential Rates of Shoreline Advance and Retreat at Coastal Barriers of Chatham and Liberty Counties, Georgia, by G. F. Oertel and C. F. Chamberlain, 1975. A technical paper that reconstructs historical patterns and rates of erosion to anticipate future erosional trends on the barrier islands between the Savannah River and the entrance to Sapelo Sound. Appears in *Transactions of the Gulf Coast Association of Geological Societies*, vol. 25, pp. 383–390. Available only at larger colleges and universities.

Wassaw Island Erosion Study, Part II: Characteristics of Water Flow at the North End of the Wassaw Barrier Island Complex, by G. F. Oertel, 1977. A technical paper that discusses water currents affecting Wassaw Sound. Technical Report Series No. 77-2 of University of Georgia, Georgia Marine Science Center, Savannah, 31406.

Sand Stabilization on the Dunes, Beach, and Shoreface of a Historically Eroding Barrier Island: Wassaw Island Erosion Study, Part III, by G. F. Oertel and J. L. Harding, 1977. This 47-page technical paper discusses erosion control experiments conducted on Wassaw Island. Sea Grant Technical Report Series No. 77-3 of University of Georgia, Georgia Marine Science Center, Savannah, 31046.

St. Catherines: An Island in Time, by David Hurst Thomas, 1988. This short, popular work was written for the general public as a study guide to accompany the videocassette of that name. A good source of current information. Published by Georgia Endowment for the Humanities (as part of the Georgia History and Culture Series), Atlanta, GA 30322.

The Cultural Resources of Blackbeard National Refuge, McIntosh County, Georgia, by Rochelle Marrinan, May 1980. The results of field and literature surveys into the archaeological and historical resources of Blackbeard (Island) National Wildlife Refuge are presented in depth. Prepared for U.S. Fish and Wildlife Service under contract to Interagency Archaeological Services, Atlanta.

Post Pleistocene Island and Inlet Adjustment Along the Georgia Coast, by G. F. Oertel, 1975. A technical paper that discusses the development of Holocene (recent) beach ridges and their effect on Georgia's tidal inlets. Includes discussions of Sapelo Sound, St. Andrew Sound, and St. Catherines Sound.

Appears in *Journal of Sedimentary Petrology*, vol. 45, pp. 150–159. Available only at larger colleges or universities.

A Sapelo Island Handbook, by Barbara Kinsey, 1982. Published by the University of Georgia Marine Institute, Sapelo Island.

Sapelo Papers: Researches in the History and Prehistory of Sapelo Island, Georgia, June 1980. Papers reviewing the patterns of historical settlement and use of Sapelo Island. Includes studies of archaeological investigations of Native American, Spanish, English, and colonial occupations. Published by West Georgia College, Carrollton, as vol. 19 of Studies in the Social Sciences series.

Sapelo Island, by B. H. Kjerfve, 1970. This article discusses marine life on Sapelo Island and includes information about Blackbeard, Cabretta, Nannygoat, and St. Simons Islands. Published in *Oceans*, vol. 3, no. 2, pp. 7–17.

Some Aspects of Modern Barrier Beach Development, Sapelo Island, Georgia, U.S.A., by J. A. Greaves, 1976. A technical paper that discusses the results of short-term monitoring of Sapelo Island's beaches, waves, currents, and weather. Published in *Pleistocene and Holocene Sediments, Sapelo Island, Georgia, and Vicinity*, Geological Society of America, Southeastern Section, Guidebook for Field Trip No. 1, April 11–13, 1966, edited by J. H. Hoyt, pp. 40–63. Available at larger colleges and universities.

Georgia Coastal Region, Sapelo Island, U.S.A., Sedimentology and Biology. VII. Conclusions, by J. D. Howard and H. E. Reineck, 1972. A technical paper that summarizes studies of the modern environments and ancient sediments of the nearshore region of the Georgia continental shelf. Published in *Senckenbergiana Maritima*, vol. 4, pp. 220–224. Available only in larger university libraries.

Pleistocene and Holocene Sediments, Sapelo Island, Georgia, and Vicinity, edited by J. H. Hoyt and others, 1966. An illustrated 78-page guidebook that contains commentaries on nine localities: Sapelo Island beach, Cabretta Island beach, Raccoon bluff, a dune ridge, a barrier spit and salt marsh, Sutherland bluff, and the Pamlico and Talbot shorelines. Guidebook published by Geological Society of America, Southeastern Section, Guidebook for Field Trip No. 1, April 11–13, 1966. Available at larger college or university libraries.

Glynn County Beach and Dune Study, by the Glynn County Beach and Dune Study Commission, 1973. Published by Georgia Department of Natural Resources, Atlanta.

Beach Erosion Control Study on Sea Island and St. Simons Island, Georgia, by the U.S. Army Corps of Engineers, 1968. Published by and available at the Savannah District Office, U.S. Army Corps of Engineers, Savannah, 31402.

Feasibility Study of Glynn County, Georgia, Beach Restoration, prepared by Olsen Associates, Inc., Coastal Engineering, December 1988. This report presents results of a feasibility study for renourishment of the shorelines of Sea Island, St. Simons Island, and Jekyll Island. The report's results, in part, are the basis for current efforts to renourish St. Simons beaches.

Brunswick Harbor, Georgia: Bar Channel, St. Simons Island Beach. Section 933 Evaluation Report and Environmental Assessment, by the U.S. Army Corps of Engineers, Savannah District, May 1990. Report regarding a recommended project for placement of beach-quality sand from advance maintenance widening of Brunswick Harbor Bar Channel onto the beach of St. Simons Island. Available at the Savannah District Office, U.S. Army Corps of Engineers, Savannah 31402.

A Field Guide to Jekyll Island, revised edition, by H. E. Taylor Schoettle, 1984. This easy-to-read field guide introduces you to the Georgia Bight and the geology and ecology of Jekyll Island, then guides you to the places where these things can be seen. Appendixes include a field guide to local plants and invertebrates, a personal safety guide, a book list, and a bibliography. An excellent field trip guide for Jekyll Island visitors and residents alike. Issued by the Georgia Sea Grant College Program, University of Georgia, Athens. Available at coastal bookstores and through Marine Extension Service offices.

Coastal Processes and Barrier Island Development, Jekyll Island, Georgia, by Vernon J. Henry and William J. Fritz, 1985. A field trip guidebook written for serious students of coastal geology. Published as the 20th Annual Field Trip Guidebook of the Georgia Geological Survey. Available in larger university libraries or through the Georgia Geologic Survey, Atlanta 30334.

The Economic Impact of Recreational Land Use in an Island Environment: A Case Study of Jekyll Island, Georgia, by C. F. Floyd and C. F. Sirmans, 1975. Compilation of data on the effect of recreational land use on employment, housing, income, migration patterns, and population. Technical Report No. SG 75-7, Georgia Marine Science Center, Savannah 31406.

Beach Erosion Control and Hurricane Protection Feasibility Report for Jekyll Island, Georgia, by the U.S. Army Corps of Engineers, Savannah District, 1974. Published by U.S. Army Corps of Engineers and available at the Savannah District Office, Savannah 31402.

The Ecology of the Cumberland Island National Seashore, Camden County, Georgia, by H. O. Hillestad and others, 1975. Inventories and describes (in 299 pp.) the natural resources of Cumberland Island National Seashore and discusses their functions and relationships. Technical Report Series No. 75-5, Georgia Marine Science Center, University of Georgia, Savannah 31406.

Site Analysis

Coastal Mapping Handbook, edited by M. Y. Ellis, 1978. A primer on coastal mapping that outlines the various types of maps, charts, and photography available; sources for such products; data and uses; state coastal mapping programs; and information appendixes and examples. A valuable starting reference for anyone interested in maps or mapping. For sale by Superintendent of Documents, U.S. Government Printing Office, Washington, DC 20402. Request publication no. 024-001-03046-2.

Floodplain Management: Ways of Estimating Wave Heights in Coastal High Hazard Areas in the Atlantic and Gulf Coast Regions, by the Federal Emergency Management Agency, 1981. This publication is of interest to planners and is available from FEMA, Washington, DC 20472.

Guidelines for Identifying Coastal High Hazard Zones, by the U.S. Army Corps of Engineers, 1975. Report outlining such zones with emphasis on "coastal special flood-hazard areas" (coastal floodplains subject to inundation by hurricane surge with a 1 percent chance of occurring in any given year). Provides technical guidelines for conducting uniform flood-insurance studies and outlines methods of obtaining 100-year-storm-surge elevations. Recommended to planners. Available from Galveston District, U.S. Army Corps of Engineers, Galveston, TX 77553.

Your Place at the Beach: A Buyer's Guide to Vacation Real Estate, edited by Sarah Friday, 1987. An excellent guide to buying vacation real estate. Although written for the North Carolina buyer, most of the information in this 28-page guide is equally applicable to the Georgia buyer. Raises the questions that you as a buyer should ask of sellers, real estate agents, government personnel, attorneys, and yourself—whether you're looking at a vacant lot, cottage, townhouse, condominium, or time-share. Available as Publication UNC-SG-87-04 from UNC Sea Grant, Box 8605, North Carolina State University, Raleigh, NC 27695-8605. Phone (919) 737-2454.

Building Construction on Shoreline Property, a checklist by C. A. Collier. Homeowners and prospective buyers of coastal property will find this pamphlet handy in evaluating location, elevation, building design and construction, utilities, and inspection. Available free from either Marine Advisory Program, G022 McCarty Hall, University of Florida, Gainesville, FL 32611, or Florida Department of Natural Resources, Bureau of Beaches and Shores, 202 Blount Street, Tallahassee, FL 32304.

Handbook: Building in the Coastal Environment, by R. T. Segrest and Associates, 1975. A well-illustrated, clearly and simply written book on Georgia coastal zone planning, construction, and selling problems. Topics include vegetation, soil, drainage, setback requirements, access, climate, and building orientation. The section on rules and regulations is dated, but the rest of the book remains timely. Includes a list of addresses for agencies and other sources of information. Available from Graphics Department, Coastal Georgia Regional Development Center, P.O. Box 1917, Brunswick, GA 31521.

Groundwater

Ground-water Data for Georgia, 1986, by J. S. Clarke and others, 1987. A report by the U.S. Geological Survey in cooperation with the Environmental Protection Division and Georgia Geologic Survey of the Georgia Department of Natural Resources. Section 2.7.4 discusses continuing groundwater depletion in the coastal area and includes maps that show the cones of depression caused by large groundwater withdrawals (mostly industrial) near Savannah, Brunswick, and St. Marys. Available in larger university libraries as USGS Open-File Report 87-376.

Ground Water: The Crisis Below, a reprint of a series of articles that appeared in the *Savannah Morning News*, October 26–31, 1981, by Betsy Neal, Kay Jackson, and Shannon Lowry. The series focuses on the growing problem of adequate water supplies in the Southeast and includes a particularly interesting article, "Low Country Is Buying Time." Available from *Savannah News-Press*, P.O. Box 1088, Savannah, GA 31402.

Ground Water in the Coastal Plains Region, A Status Report and Handbook, compiled by A. D. Park, 1979. This report was prepared for the Coastal Plains Regional Commission and addresses the subject of groundwater for the five Southeastern states. Although specific Georgia coastal groundwater problems are not presented, a good summary (but now dated) of the state's programs and needs are noted. Other states' problems and programs provide a basis for comparison. Includes an extensive list of references, including studies on saltwater intrusion, aquifer depletion, and water contamination associated with waste disposal. The appendix provides an extensive list of groundwater agencies for the states. Available from Coastal Plains Regional Commission, 215 East Bay Street, Charleston, SC 29401.

Planning and Management

Coastal Management: Planning on the Edge, edited by D.R. Godschalk and Kathryn Cousins, 1985. This special issue of the *Journal of the American Planning Association* (vol. 51, no. 3) focuses on coastal management issues. Of particular interest is the story of Nags Head, N.C., a community which has chosen relocation of buildings instead of shoreline stabilization. Highly recommended reading. Available through larger college and university libraries.

A Demographic Study of the Coastal Counties, 1790–1980, by Davor Jedlicka, 1981. Investigates the evolution of coastal Georgia's population growth and associated land use and resource use as indicated by changes in the local economies (such as agriculture and fishing). Published by Georgia Marine Science Center, University of Georgia, as Technical Report No. 81-3.

Coastal Growth and Development: Issues and Alternatives. Published by Coastal Area Planning and Development Commission, 1984. Available from Coastal Georgia Regional Development Center, P.O. Box 1917, Brunswick, GA 31521.

Coastal Area Development Profile, Fiscal Year Overview. Published annually by the Coastal Area Planning and Development Commission and available from Coastal Georgia Regional Development Center, P.O. Box 1917, Brunswick, GA 31521.

Conditions, Trends, Issues in Coastal Georgia, by the Coastal Area Planning and Development Commission, 1985. Available from Coastal Georgia Regional Development Center, P.O. Box 1917, Brunswick, GA 31521.

A Resource Planning Process for Georgia's Coast, by L. F. Dean, 1975. Discusses management programs for the Georgia coastal zone. Article on p. 107 in *The Value and Vulnerability of Coastal Resources*, published by Georgia Department of Natural Resources, Atlanta.

The Water's Edge: Critical Problems of the Coastal Zone, edited by B. H. Ketchum, 1972. Scientific summary of coastal zone problems. Published by the MIT Press, Cambridge, MA 02139.

Design with Nature, by Ian McHarg, 1969. A now-classic text on the environment. Stresses that when man interacts with nature, he must recognize its processes and governing laws and realize that it presents both opportunities and limits on human use. Published by Doubleday and Company, Garden City, NY 11530.

And Two If By Sea: Fighting the Attack on

America's Coasts, by Beth Milleman, 1986. Handy 109-page reference—thorough and concise—to the threats facing our coasts nationwide. Of special interest is the chapter on "Coastal Hazards." Appendixes include lists of state coastal management program offices (Georgia is not included since the state does not participate in the federal coastal zone management program) and environmental groups concerned with coastal issues. Available from Coast Alliance, 218 D Street, S.E., Washington, DC 20003. Phone (202) 466-5045.

Barrier Islands, by H. C. Miller, 1981. Excellent review of how the federal government has stimulated barrier island development with resulting financial losses in tax dollars. Published in *Environment*, vol. 23, no. 9, pp. 6–11, 36–42.

Ecological Determinants of Coastal Area Management (2 vols.), by Francis Parker, David Brower, and others, 1976. Volume 1 defines barrier island and related lagoon estuary systems and the natural processes that operate within them. Outlines man's disturbing influences on coastal environments and suggests management tools and techniques. Volume 2 consists of appendixes that include information on coastal-ecological systems, man's impact on shorelines, and tools and techniques for coastal area management. Also contains a good bibliography. For sale by Center for Urban and Regional Studies, University of North Carolina, 108 Battle Lane, Chapel Hill, NorthCarolina 27514.

Coastal Environmental Management, prepared by the Conservation Foundation, 1980. Guidelines for conservation of resources and protection against storm hazards, including ecological description and management suggestions for coastal uplands, floodplains, wetlands, banks and bluffs, dunelands, and beaches. Part II presents a complete list of federal agencies and their authority under law to regulate coastal zone activities. A good reference for planners and people interested in wise land management. Available from Superintendent of Documents, U.S. Government Printing Office, Washington, DC 20402.

Development of the Coast: Facing the Tough Issues. These published final proceedings of the Coastal States Organization conference held in Charleston in 1979 give an abbreviated overview of the wide range of problems generated by coastal development. Available from CSO, Conference Management Associates, Ltd., 1044 National Press Building, Washington, DC 20045.

The Fiscal Impact of Residential and Commercial Development, A Case Study, by T. Muller and G. Dawson, 1972. A classic study which demonstrates that development may ultimately increase, rather than decrease, community taxes. Available from the Publications Office, Urban Institute, 2100 M Street, N.W., Washington, DC 20037. Refer to URI-22000 when ordering.

Environmental Improvement Through Economic Incentives, by Frederick Anderson and others, 1977. A good introduction (195 pp.) on using economic incentives to promote the improvement of environmental quality. Although some ideas will be new to much of the general public in the United States, they deserve careful consideration and vigorous discussion. Published for Resources for the Future by Johns Hopkins University Press, 701 W. 40th Street, Suite 275, Baltimore 21211. Available in many libraries.

Report of the Conference on Marine Resources of the Coastal Plains States, 1974. Collection of papers presented at a meeting in Wilmington, N.C. Topics include seabed mineral resources, sport fishing, recreation and tourism, and coastal zone planning. Of special interest is the paper "Responsible Development and Reasonable Conservation," by David Stick. Sponsored and published by Coastal Plains Center for Marine Development Services. Available in some university libraries.

Coastal Ecosystem Management, by John Clark, 1977. This 928-page text covers most aspects of the coastal zone from descriptions of processes and environments to legal controls and outlines for management programs. Essential reading for planners and beach community managers. Published by John Wiley and Sons, New York. Available in most university libraries.

Coastal Affair, edited by Jennifer Miller, May/June 1982. A special subject issue of *Southern Exposure*, vol. 10, no. 3. Explores a wide range of coastal issues: physical, social, and economic. The details given are somewhat dated, but the issues remain much the same. Available from *Southern Exposure*, P.O. Box 531, Durham, NC 27702.

Storm on the Horizon: The National Flood Insurance Program and America's Coasts, by B. Milleman, 1989. Discusses the history, aims, and effects of the National Flood Insurance Program, the country's second largest domestic program (second only to the Social Security system). Available form the Coast Alliance, 235 Pennsylvania Avenue S.E., Washington, DC 20003. Phone (202) 546-9554.

Using Common Sense to Protect the Coasts: The Need to Expand the Coastal Barrier Resources System, produced and distributed by the Coast Alliance, 1990. Provides a general introduction to coastal barriers, discusses federal subsidies for their development, and explains the aims of the Coastal Barrier Resources Acts. Available from the Coast Alliance, 235 Pennsylvania Avenue S.E., Washington, DC 20003. Phone (202) 546-9554.

Regulations

Answers to Questions About the National Flood Insurance Program, by FEMA, 1990, FIA-2. Pamphlet explaining basics of flood insurance and providing phone numbers and addresses of FEMA offices. Related publications include *Guide to Flood Insurance Rate Maps* (FEMA-14) and *Appeals, Revisions, and Amendments to Flood Insurance Rate Maps* (FIA-12). Request by title and number, free from Federal Emergency Management Agency, P.O. Box 70274, Washington, DC 20024.

Ocean and Coastal Law, by R. Hildreth and R. Johnson, 1983. Discusses (1) problems posed by earlier nonmanagement of public resources and the gradual public awakening to the need for a comprehensive legal framework for the coastal zone, (2) ownership and boundary questions, (3) state common law, (4) offshore issues, (5) alteration of waterways and wetlands, and (6) federal and state coastal zone management programs. For anyone with a serious interest in legal issues. Published by Prentice-Hall, Englewood Cliffs, N.J. 07632.

Federal Legislation

The following are specific references to the federal legislation mentioned in chapter 6. The first reference under each act shows where it can be found in the *United States Code*. The second indicates its place in *Statutes at Large*.

National Flood Insurance Act of 1968 (P.L. 90-448), enacted on August 1, 1968.
(1) 42 U.S.C. sect. 4001 et seq. (1976).
(2) 82 Stat. 476, Title 13.

Federal Water Pollution Control Act Amendments of 1972 (P.L. 92-500), enacted on October 18, 1972.
(1) 33 U.S.C. sect. 1251 et seq. (1976).
(2) 86 Stat. 816.

Marine Protection, Research and Sanctuaries Act of 1972 (P.L. 92-532), enacted on October 23, 1972.
(1) 33 U.S.C. sect. 1404 et seq. (1976).
(2) 86 Stat. 1052.

Flood Disaster Protection Act of 1973 (P.L. 92-234), enacted on December 31, 1973.
(1) 42 U.S.C. sect. 4001 et seq. (1976).
(2) 87 Stat. 975.

Water Resources Development Act of 1974 (P.L. 93-251), enacted on March 7, 1974.
(1) 16 U.S.C. sects. 4601-13, 4601-14, 460ee (1976); 22 U.S.C. sect. 275a (1976); 33 U.S.C. sects. 59c-2, 59k, 579, 701b-11, 701g, 701n, 701r, 701r-1, 701s, 709a, 1252a, 1293a (1976); 42 U.S.C. sects. 1962d-5c, 1962d-15, 1962d-16, 1962d-17 (1976).
(2) 88 Stat. 13, Title 1.

Water Resources Development Act of 1986 (P.L. 99-662), passed November 17, 1986. Available as House Report 99-1013, 99th Congress, 2nd Session, from Senate Documents Room, Hart Senate Office, Room B-04, Washington, DC 20510.
(1) 33 U.S.C. sect. 2201.
(2) 100 Stat. 4082.

Building or Improving a Home

Both current and prospective owners and builders of homes in hurricane-prone areas should be familiar with the principles of hurricane-resistant construction. The references listed contain sound information that should help residents minimize losses caused by extreme wind or rising water. Many of these publications are free. The government publications are paid for by your taxes. Why not use them?

Coastal Design: A Guide for Builders, Planners, and Homeowners, by Orrin H. Pilkey, Jr., Orrin H. Pilkey, Sr., Walter D. Pilkey, and W. J. Neal, 1983. A detailed companion volume and construction guide to accompany the Living with the Shore series. Includes types of shoreline, individual residence construction, methods for making older structures stormworthy, high-rise buildings, mobile homes, coastal regulations, and the future of the coastal zone. Published by Van Nostrand-Reinhold Company, New York.

Floodproofing Non-Residential Structures, prepared by Booker Associates, Inc., for FEMA, 1986. A 199-page illustrated guide to floodproofing buildings. Recommended reading for planners, community officials, managers, and owners of all types of buildings located in flood zones, both coastal and riverine. Available as Publication No. 1986-621-393/00128 from Superintendent of Documents, U.S. Government Printing Office, Washington, DC 20402.

Flood Emergency and Residential Repair Handbook, by FEMA, March 1986. Publication FIA-13. FEMA Publications, P.O. Box 70274, Washington, DC 20024.

FEMA Publications Catalog, by FEMA. Lists more than 300 publications available from that agency to assist everyone from individual property owners to emergency managers. Request from FEMA Publications, P.O. Box 70274, Washington, DC 20024.

A Coastal Homeowner's Guide to Floodproofing, by the Massachusetts Disaster Recovery Team. Booklet combines a checklist to take homeowners through the process of floodproofing existing houses and tips on dealing with engineers and contractors. Although prepared for New England, the guide is appropriate to some Georgia construction. Available from Office of the Lieutenant Governor, State House, Boston, MA 02133.

Standard Building Code, 1979. (Previously known as the *Southern Standard Building Code* and still frequently referred to by that name). Available from Southern Building Code Congress International, 900 Mountclair Road, Birmingham, AL 35213.

Dwelling House Construction Pamphlet of Southern Standard Building Code. Conforms to the Southern Standard Building Code and applies only to dwellings of wood-stud or masonry-wall construction. Available from Southern Building Code Congress International (above) or Southern Building Code Publishing Company, 900 Mountclair Road, Birmingham, AL 35213.

The Uniform Building Code. Available from International Conference of Building Officials, 5360 South Workman Mill Road, Whittier, CA 90601.

The BOCA Building Code. Available from Building Officials and Code Administrators International, Inc., 17926 South Halsted Street, Homewood, IL 60430.

Coastal Construction Building Code Guidelines, edited by R. R. Clark, 1980. Although developed for Florida, these guidelines are applicable to other coastal areas and make specific recommendations to strengthen the Standard Building Code in those areas. Available as Technical Report No. 80-1 from Division of Beaches and Shores, Florida Department of Natural Resources, 3900 Commonwealth Boulevard, Tallahassee, FL 32303.

Hurricane Exposes Structure Flaws, by Herbert S. Saffir, 1971. In *Civil Engineering Magazine*, February 1971, pp. 54–55. Available from most university libraries.

Potential Wind Damage Reduction Through the Use of Wind-Resistant Building Standards. Of interest to builders of new structures. Available from Texas Coastal and Marine Council, P.O. Box 13407, Austin, TX 78711.

Model Minimum Hurricane-Resistant Building Standards for the Texas Gulf Coast. Although written specifically for Texas, this publication will interest all coastal residents and property owners. Available from Texas Coastal and Marine Council, P.O. Box 13407, Austin, TX 78711.

Estimating Increased Building Costs Resulting from Use of a Hurricane-Resistant Building Code. Of interest to builders of new structures. Available from Texas Coastal and Marine Council, P.O. Box 13407, Austin, TX 78711.

Manufactured Home Installation in Flood

Hazard Areas, prepared by the National Conference of States on Building Codes and Standards, Inc., for FEMA, 1985. A 110-page guide to design, installation, and general characteristics of manufactured homes with respect to coastal and flood hazards. Anyone contemplating buying such a structure, or already living in one, should read this publication and follow its suggestions to lessen potential losses from flood, wind, and fire. Available as publication no. 1985-529-684/31054 from Superintendent of Documents, U.S. Government Printing Office, Washington, DC 20402. Request by title and number.

Structural Failures: Modes, Causes, Responsibilities, 1973. See especially the chapter "Failure of Structures Due to Extreme Winds," pp. 49–77. Available from Research Council on Performance of Structures, American Society of Civil Engineers, 345 East 47th Street, New York, NY 10017.

Wind-Resistant Design Concepts for Residences, by Delbart B. Ward. Displays with vivid sketches and illustrations construction problems and methods of tying structures to the ground. Extensive text and excellent illustrations devoted to methods of strengthening residences. Offers recommendations for relatively inexpensive modifications that will increase the safety of residences subject to severe winds. Chapter 8, "How to Calculate Wind Forces and Design Wind-Resistant Structures," should particularly interest designers. Available as TR-83 from Civil Defense Preparedness Agency, Department of Defense, The Pentagon, Washington, DC 20301, or from Civil Defense Preparedness Agency, 2800 Eastern Boulevard, Baltimore, MD 21220.

Interim Guidelines for Building Occupant Protection from Tornadoes and Extreme Winds, TR-83A, and *Tornado Protection—Selecting and Designing Safe Areas in a Building*, TR-83B. These publications supplement *Wind-Resistant Design Concepts for Residences* and are available from the same addresses.

Hurricane-Resistant Construction for Homes, by T. L. Walton, Jr., 1976. An excellent booklet produced for residents of Florida, but equally useful to those on the Georgia coast. A good summary of hurricanes, storm surge, damage assessment, and guidelines for hurricane-resistant construction. Presents technical concepts on probability and its implications for home design in hazard areas. Available from Florida Sea Grant Publications, G022 McCarty Hall, University of Florida, Gainesville, FL 32611.

Coastal Construction Manual, prepared by Dames & Moore and Bliss & Nyitray, Inc., for the Federal Emergency Management Agency, February, 1986. A guide to the coastal environment with recommendations on site and structure design relative to the National Flood Insurance Program. Includes discussions of the National Flood Insurance Program, building codes, coastal environments, site and structure design recommendations, and examples. This second edition also includes a new chapter on the design of large structures at the coast. Appendixes include design tables, bracing lexamples, design worksheets, equations and procedures, construction costs, and a list of references. Additions to this edition include computer program listings and a sample construction code for use by coastal municipalities. Available from Superintendent of Documents, U.S. Government Printing Office, Washington, DC 20402. Ask for FEMA publication no. 55. Also available from FEMA offices.

Elevated Residential Structures, prepared by the American Institute of Architects Foundation (1735 New York Avenue, N.W., Washington, DC 20006) for the Federal Emergency Management Agency, 1984. An excellent outline of coastal and riverine flood hazards and the necessity for proper planning and construction. Discusses the National Flood Insurance Program, site analysis and design, design examples, and construction techniques; includes illustrations, glossary, references, and worksheets for estimating building costs. Available as stock no. 1984-0-438-116 from Superintendent of Documents, U.S. Government Printing Office, Washington, DC 20402. Also available as FEMA publication no. 54 from FEMA offices.

Design Guidelines for Flood Damage Reduction, prepared by the American Institute of Architects for the Federal Emergency Management Agency, 1981. This is the companion volume to the previous two references and is available from the same sources. Ask for FEMA publication no. 15.

Flood Emergency and Residential Repair Handbook, prepared by the National Association of Homebuilders Research Advisory Board of the National Academy of Science, 1980. Guide to floodproofing as well as step-by-step cleanup procedures and repairs, including household goods and appliances. Available from Superintendent of Documents, U.S. Government Printing Office, Washington, DC 20402. Order stock no. 023-000-00552-2.

Your Home Septic System, Success or Failure? Brochure with answers to commonly asked questions on home septic systems. Available from UNC Sea Grant, 1235 Burlington Laboratories, North Carolina State University, Raleigh, N.C. 27607.

Wood Structures

Houses Can Resist Hurricanes, by the U.S. Forest Service, 1965. An excellent paper with numerous details on general construction. Pole-house construction is treated in detail (pp. 29–45). Available as *Research Paper FPL 33* from Forest Products Laboratory, Forest Service, U.S. Department of Agriculture, P.O. Box 5130, Madison, WI 53705.

Wood Structures Survive Hurricane Camille's Winds. Available as *Research Paper FPL 123*, October 1969, from Forest Products Laboratory, Forest Service, U.S. Department of Agriculture, P.O. Box 5130, Madison, WI 53705.

Wood Structures Can Resist Hurricanes, by Gerald E. Sherward, 1972. See *Civil Engineering Magazine*, September 1972, pp. 91–94. Available from most university libraries.

Masonry Construction

Standard Details for One-Story Concrete Block Residences, by the Masonry Institute of America. Contains nine foldout drawings that illustrate details of constructing a concrete-block house. Presents principles of reinforcement and good connections aimed at design for seismic zones, but these apply to design in hurricane zones as well. Written for both layman and designer. Available as Publication 701 from Masonry Institute of America, 2550 Beverly Boulevard, Los Angeles, CA 90057.

Masonry Design Manual, by the Masonry Institute of America. A 384-page manual that covers all types of masonry, including brick, concrete block, glazed structural units, stone, and veneer. Comprehensively detailed and well-presented. This book is probably of more interest to designers than laymen. Available as Publication 601 from Masonry Institute of America, 2550 Beverly Boulevard, Los Angeles, CA 90057.

Pole-House Construction

Pole House Construction and Pole Building Design. Available from American Wood Preservers Institute, 1651 Old Meadows Road, McLean, VA 22101.

Mobile Homes

Protecting Mobile Homes from High Winds, TR-75, prepared by the Civil Defense Preparedness Agency, 1974. This out-of-print booklet outlines methods for tying down mobile homes and means of protection such as positioning and windbreaks.

Suggested Technical Requirements for Mobile Home Tie Down Ordinances, TR-73-1, prepared by the Civil Defense Preparedness Agency, July 1974. Should be used in conjunction with TR-75, *Protecting Mobile Homes from High Winds*, described above. Available from U.S. Army Publication Center, Civil Defense Branch, 2800 Eastern Boulevard (Middle River), Baltimore, MD 21220.

Appendix E Selected Field Stops on Tybee Island

An excellent field-oriented introduction to natural coastal processes has been written by Taylor Schoettle for Jekyll Island. (The full citation is given in appendix D.) Schoettle also has written a field guide for Sea Island, but that island is accessible to fewer people. We strongly encourage the reader to drive through the enjoyable trip outlined for Jekyll Island.

This appendix includes a brief introduction to a few selected stops in the Tybee Island area. The aim is to provide an on-site look at some of the processes and events discussed in the preceding text.

The field trip begins in Savannah, where Victory Drive crosses the Wilmington River, and ends on Tybee Island.

*0.0 miles: Highway 80/Victory Drive, at foot of Wilmington River Bridge**

This high-span bridge over the Intracoastal Waterway was built to replace an aging drawbridge, a hazard to efficient evacuation. To the south (right), the town of Thunderbolt lines the river bluff. Shrimp boats share the waterfront with sailboats and yachts, and seafood distributors, restaurants, houses, and condominiums vie for upland space. Contrasting resource uses are immediately evident, but conflicts among the resource users are only hinted at.

---

*Vehicle mileage is given from the starting point at Highway 80 and Victory Drive. The numbers at various stops are cumulative.

Thunderbolt is undergoing a transition from a shrimping community, with an economy based on water-related industries, to a bedroom community for Savannah. The number of family-operated fleets of shrimp boats has dramatically decreased since the early 1970s, and recreational and residential use of the waterfront area is becoming dominant.

In the midst of the marshes across from the bluff at Thunderbolt you'll see an area of dry land with scrub vegetation. This "upland" is a dredge spoil site. When sediment is dredged from the river navigational channel, it is dumped onto these spoil sites.

1.6 miles: Intersection of Highway 80 and Highway 367
Turn right onto Highway 367/Johnny Mercer Boulevard.

The highway built to Tybee in 1933 created this route through Whitmarsh, Talahi, and Wilmington Islands. Although the islands are now surrounded by marshes, far from the open ocean, 25,000 to 40,000 years ago they were ocean barriers. Within recent centuries the islands have evolved from relatively undisturbed wilderness, to agricultural areas and scattered resorts, to developing suburbs of Savannah.

Although most of the population growth initially occurred on Wilmington, the islands of Whitmarsh and Talahi are experiencing an increase in residents and demands for services. The relatively large size of the islands will accommodate some population growth; however, the rate and type of growth have not been planned or managed well. Demands for basic services such as adequate road systems, drinking water, sewage treatment, and schools have created strains on limited resources and services.

Condominiums and businesses on Whitmarsh and Talahi began to proliferate during the mid-1980s. As you drive along Highway 367, note the number of new housing developments. The problems associated with this rapid growth on Wilmington are likely to occur on Whitmarsh and Talahi Islands.

As you approach the bridge over Turner's Creek, notice the dense stands of tall rushlike spikes of green and brown grass near the islands. This plant is *Juncus*, the landmark of high marsh. These sections of the salt marsh are flooded by tides less frequently than the lower portions. Clusters of red cedar trees are scattered along the upper edges of the *Juncus* marsh. The red cedar, *Juniperus virginiana*, is often found at Native American midden sites, which usually consist of a mound of oyster shells. Weathering of the oyster shells makes the surrounding soil more alkaline—a condition that suits the red cedar's requirements.

3.7 miles: Bridge Over Turner's Creek to Wilmington Island
As you cross Turner's Creek bridge, take advantage of the fantastic view. To the south (right) the creek winds along a Wilmington

Island bluff. As in Thunderbolt, the varied uses of coastal resources are well-illustrated by the many different waterfront facilities.

Looking to the north (left) from atop the bridge, you'll see Turner's Creek winding through the *Spartina* salt marshes that separate Whitmarsh and Wilmington Islands. In the distance the U.S. Highway 80 bridge spans the wetlands between Whitmarsh and Talahi Islands. For one moment, imagine the scene if storm waters filled the marshes and flooded the islands....

The storm surge of the "great hurricane" of August 27, 1893, flooded all of the islands—Tybee, Little Tybee, Wilmington, Talahi, Whitmarsh, Oatland, Skidaway—between the mainland shore and the ocean! The 17.5-foot storm surge that completely inundated Tybee did the same on these former barriers, although they are miles from the sea. And according to FEMA's flood hazard maps, virtually all of these islands will be under water again during the surge from a similar 100-year storm.

5.7 miles: Former Landfill/Recreation Area Entrance

A landfill for disposal of solid waste covers several acres to the right. A recreation area has been constructed on areas of the site, now filled to capacity.

Disposal of solid waste poses a major problem for any island community. Lack of available space, potential contamination of shallow aquifer water supplies, and high costs of waste removal are important considerations in solid waste management.

6.5 miles: Intersection of Highway 367 with U.S. Highway 80

Turn right onto Highway 80 from Highway 367. The salt marshes to the right of the Bull River bridge are part of the widest expanse of salt marsh on the Georgia coast. These marshes were formed and influenced by the Savannah River delta. The mud bottoms of the shallow tidal creeks are often exposed at low tide, revealing huge oyster clusters. During extremely high tides the marsh grasses may be completely covered.

When extremely high spring tides occur during October and November the birds known as clapper rails often are forced from their homes in the marsh grass. In autumn such tides are called "marsh hen tides" because the rails are hunted most easily at these times.

Tybee Island is 9.5 miles to the east of the bridge. Occasionally, the causeway and the road to Tybee (fig. 5.7) are flooded by storm tides or spring high tides of 8 to 10 feet. Although a mere inconvenience during spring tides, flooding of U.S. Highway 80 may pose serious evacuation problems during a severe storm or hurricane.

Along the rest of the drive to Tybee, the road parallels the former railway that ran between the mainland and the island from 1877 to 1933. You can spot the old railway bed to the north (left) by the shrubs and trees now growing on this relatively high ground, which also serves as habitat for yaupon hollies, myrtles, yuccas, diamondback rattlesnakes, and marsh rabbits.

10.9 miles: Fort Pulaski National Monument

Fort Pulaski is located on Cockspur Island, which separates the north and south channels of the Savannah River. Since the time when Georgia was established as a colony, fortifications on this site have guarded the river's entrance. A walk around the parapet of Fort Pulaski provides great views of the river channels, of Tybee's north end, and of the extensive salt marshes. The rock jetties marking the boundaries of the Savannah River's north channel are visible at low tide.

A tour of Cockspur Island offers a close view of a dredge spoil island. As you cross the south channel bridge to Cockspur, the only watercraft you see are likely to be small recreational vessels or shrimp boats. The river's south channel was the major shipping passage to Savannah's port until the mid-1870s. In 1876 dredging of the north channel for use by large, deep-draft ships was begun, and to this day it serves as the main shipping channel. You can get a good view of the north channel by walking the trail at the end of the parking area. Fort Green, destroyed during the hurricane of 1804, was located on Cockspur Island.

10.9 miles: Intersection of Fort Pulaski Entrance Road and Highway 80

As you continue along the causeway toward Tybee Island, realize that you're driving over the sand of dunes formerly covered by sea oats. When work on the causeway project began, fill material for its construction was taken from dunes along the northern ocean beach.

A heron rookery (nesting area) once existed near here at a small pond, but was destroyed by the road's construction. In years past the city of Savannah Beach used the land between the causeway and the river as a solid-waste disposal site. After the dump site closed, plans were announced for constructing condominiums, but these plans have not materialized.

As you continue, the road nears two Holocene-age hammocks (land unusually higher than its surroundings); one is marked by a sign at Spanish Hammock. Another series of parallel ridges of Holocene-age hammocks is buried, partially or entirely, by the salt marsh.

An area of marshland adjacent to the hammock along Chimney Creek was filled with dredge spoil to create upland. Construction of mobile homes, houses, and finger canals soon followed.

As you come onto Tybee Island, turn left onto Byers Street. Turn right onto Bay Street.

Rivershore Access Just Beyond McKenzie Street

Follow the path to the river beach. The condominiums to the west (left) are located along the last section of river beach on Tybee's northwestern edge. Less than half a mile to the west (left) the beach ends at a tidal creek beyond which salt marsh extends to Lazaretto Creek. The beach fronting these condominiums experienced net accretion between 1866 and 1974. However, from 1925 through 1974 this portion of the island experienced net erosion.

Note the present condition of the beach and dune system. Are the dunes truncated? How well-developed is the system of foredunes and backdunes? The vegetation between dunes, including bayberry with its aromatic dark-green leaves, indicates that this is not a young system.

A walk along the beach to the west (left) may show evidence of beach and dune erosion. Take notice of any crude, individual efforts at erosion control—discarded Christmas trees and debris along the upper beach, for instance.

Continue driving northeast on Bay Street. Follow Bay Street until it ends. Turn right on Polk Street and follow it to Taylor Street, the first street on the left. Turn left on Taylor Street and drive along the edge of the River Oaks Campground.

The relatively undisturbed assemblage of live oaks, a section of a former maritime forest, is the largest group of such trees you're likely to see on this tour. During the drive around the island, remember that Tybee, prior to the arrival of the English, was forested over much of its upland. Forests were removed for residences, livestock, pasture, timber, and military purposes.

From Taylor Street, turn left onto Van Horne Street.

You have entered the area of the former Fort Screven Army Reservation. Former barracks and the train/trolley station are located along Van Horne Street. The northern end of Tybee has served as a military post since colonial days because of its excellent vantage point at the Savannah River entrance. The Fort Screven Coast Artillery Base, established in 1897, was decommissioned and opened to residential use after World War II.

From Van Horne Street, turn right onto Hollywood Avenue. Enter the parking area of Lighthouse Point Condominiums. Visitor parking is located near the swimming pool. (Note: Only condominium residents, visitors of residents, or persons granted special permission may enter the parking area.)

Lighthouse Point Condominiums

Walkways over the sand dunes lead to the north end river beach. Walkovers like these are becoming increasingly common as beach dwellers become more sensitive to the benefits of protecting dunes from foot traffic.

A line of wooden pilings perpendicular to and under the walkway are remnants of a former bulkhead constructed at the pre-1975 mean high water line. This portion of Tybee has a history of accretion and erosion cycles (fig. 5.9), with considerable accretion since the mid-1970s.

Where did this new bulge of sand come from

(fig. 5.10)? When the Corps of Engineers executed the first phase of the Tybee beach replenishment project in 1976, no sand was placed along this portion of the river beach. However, the terminal groin on the island's north end began to settle, sinking more deeply than expected. Sand from the renourished beach immediately south of the groin was transported over the groin and around to the northwest to this beach. Within three years this stretch had accreted rapidly—so rapidly, in fact, that a channel marker was soon stranded by sands from the new beach.

With the new expanse of beach, the site of the condominiums was better protected from storms. Yet this portion of the island has a long-term history of erosion, especially from 1925 to 1974. With the height of the north end terminal groin raised (in 1987) to stop the artificial beach's loss of sand, what will the effects be on this part of the beach? If erosion occurs, will the condominiums be affected?

Since 1866 Tybee's northeastern tip has been the island's most rapidly changing part. Figure 5.9 illustrates the sediment loss from the northeast ocean beach as a northerly spit built up and moved to the northwest over a period of 31 years. At one time a wharf extended into the Savannah River from the Army reservation in this area. The wharf and other structures built to take advantage of the newly created spit were lost when the spit eroded from 1931 to 1974. Only after the 1976 renourishment did this area begin to grow again as sand from the nourished beach escaped to the river beach. The present river beach built up at rates of 12.5 to 15 feet per year.

As you leave the parking lot, turn left onto Hollywood Avenue. Follow the road as it becomes Pulaski Avenue and winds between former barracks and gun batteries. When Pulaski intersects Taylor Street, turn left. Turn right from Taylor Street onto Meddin Drive.

Tybee Lighthouse

An incredible bird's-eye view of Tybee is available from the lighthouse's observation balcony. An entrance fee is charged, but the view is worth the price. Shoreline position, erosion control structures, patterns of land use, and island infrastructure can be easily observed.

A view from the lighthouse or from the observation platform at the Tybee Museum allows you to imagine the same view in 1867 when the shoreline was about one-half mile to the east of today's shoreline. (Refer to Figure 5.10 for net MHW changes.) Gun batteries, parade grounds, and a fort observation platform disappeared as the shoreline receded more than 1,500 feet from 1867 to 1931. Another 1,300 feet have receded since 1931.

U.S. Army engineers developed and tested on these north end beaches many of the groins that have been built on Tybee. Now useless wooden groins from the 1940s and the large rock groin built in 1975 are visible from the lighthouse.

The first seawall on Tybee was constructed along these northern shores in 1907. Notice the terminus of the 1941 seawall and supporting rock revetments near the southern end of the parking area near the beach. The accretion to the north of the Lighthouse Point condominiums is easily seen from the observation platform.

Look for the highest inland points on Tybee. The high ridges parallel to the river beach, Officers' Row with its line of chimney-topped houses, and the center of the island's south end are some highlights. Notice the parallel ridges behind the island; these indicate former shorelines. Finally, observe the condition of the beaches and dunes along the shoreline to the south. Can you locate areas of erosion in contrast to more stable shoreline areas?

Instead of driving, you may walk to the next beach location, which is behind the Tybee Museum.

15.1 miles: Intersection of Meddin Drive and Gulick Drive
Turn left from Meddin Drive onto Gulick Drive to enter the north end parking area. During summer months a fee is charged for the parking area.

Tybee Museum

If you take the Tybee Museum tour, spend a moment looking at the old charts and maps of the area. On the plat maps by Percey Seyden, you'll see that some of the island's properties are actually on the beach or under water. Sometimes residents will continue paying taxes, even on submerged lands, in the hopes that nature may one day see fit to return some subaerial sand to their plots. In particular, notice the photograph of the Martello Tower (fort observation tower).

From the museum's observation platform, you'll have an excellent view of the beach and the island's interior. Compare the view of the present island with that from the old photographs of the Fort Screven parade ground area. Referring to the viewing platform chart, pay attention to changes in the shoreline's position since 1866.

Walk to the beach on the dune trails at the northeastern edge of the parking area. A restoration project initiated in 1988 included construction of the walkways and creation of dunes by putting up sand fencing and planting sea oats. Foot traffic destroys dunes and inhibits vegetation growth. The construction of these dune crossings is an impressive and progressive action in shoreline management.

From atop a dune crossing, look to the south (right) along the shoreline. You'll see the condition of the shoreline and the status of development adjacent to the beach. Immediately to the south are the houses of Officers' Row. A 1989 rezoning of this area by the Tybee City Council (against the recommendations of the island's planning advisors) opened the door for building residential units here, seaward of Officers' Row.

To the north (left) is the 1975 terminal stone groin, the Savannah River, and South Carolina's Daufuskie and Hilton Head Islands. You may be fortunate enough to sight a ship passing through the 38-foot channel, which extends 7 miles offshore. Many geologists believe that erosion trends on Tybee are linked to the navigation channel dredging.

Exit the parking area and turn left onto Meddin Drive. At Cedarwood Avenue, turn left. Turn left on Van Horne Street as you pass the Oceanview Nursing Home. Horn Street winds to the south and straightens into Second Avenue. At the intersection of Second Avenue and Highway 80 (First Street), turn left. After the road turns to the right, Highway 80 is called Butler Avenue.

Intersection of Butler Avenue and Third Street

Park either on Butler Avenue or on Third Street and walk to the beach in the vicinity of the condominiums.

These condos were constructed in the late 1970s after the Corps of Engineers' beach replenishment project was completed (1976). Before the artificial beach was created, this portion of the island's upland suffered from severe wave attack. After the damage caused by Hurricane Dora in 1964, the seawall along the shoreline was reinforced with granite revetment. After the 1976 beach restoration an attractive, wide beach stretched in front of the condominiums, the first ones to be built adjacent to Tybee's ocean beach. Condo owners purchased units when a lovely beach stretched from their doors to the ocean.

Soon, however, the ocean waters were flowing under their doors and under the condominiums when storm waves battered and washed over the exposed, revetment-protected seawall. Within months after beach replenishment was completed, the sands fronting the condominiums had severely eroded. The beach level dropped several feet, leaving the stairs to the beach from the condominium suspended in midair.

Unfortunately, the original owners of the condo units were not aware of the history of erosion along this portion of the shoreline (fig. 5.11).

While standing in front of the condos, notice the beach areas to the north and south. The 1976 beach disappeared quite rapidly from this area. How is the 1987 artificial beach faring (fig. 5.17)?

The counterclockwise rotation and recession of Tybee's northern shoreline since 1866 (fig. 5.9) has pivoted on a nodal point located between Second and Third Streets. This area may continue to experience similar erosion patterns.

Tidal and wave currents are largely responsible for sediment transport along the beach. A 1974 study by George Oertel indicated that during the incoming tide, tidal currents diverge in opposite directions near Ninth Street. A northward-flowing current was generated

Appendix E **183**

during the incoming tide to the north of Ninth Street, while southward movement of sand occurred to the south. (The regional annual net movement of sand remains north to south.) Other studies have identified the Sixth Street area as a divergence point for wave-transported sediment.

Return to your car and continue driving south along Butler Avenue.

Intersection of Butler Avenue and Tenth Street

This central portion of Tybee's beach has remained relatively stable over the years since records of shoreline position have been kept (fig. 5.9). The stretch between Seventh and Fourteenth Streets provides attractive recreational space (fig. 5.12). Unfortunately, the only nearby parking spaces are located on these residential streets and on Butler Avenue. Some spaces have been eliminated at the request of property owners through the creation of No Parking zones.

The dune crossing on this street was erected in the late 1970s. Prior to the 1976 beach restoration a well-developed dune system did not exist here. After the replenishment, a set of primary dunes formed and the backdunes dramatically increased in size. (The 1941 seawall underlies the backdunes today; however, the cap of the wall was exposed in the early 1970s.) Compare the interdune area here with that which fronts Officers' Row on Tybee's north end.

The existence of this relatively stable portion of island shoreline is fortunate for residents near these beaches. Although storm wave overwash may occasionally bring some sand, debris, and water into the interdune area, the dune system provides excellent flood protection for the adjacent upland property. As explained in chapter 4, although Tybee beaches may lose significant amounts of sand after renourishment projects, the beach "life" here seems to last longer than elsewhere on the island.

As you return to your car, you can see how much development has taken place between the beach and Butler Avenue. Formerly, an undeveloped strip of land lay between the beach and any houses; however, as the island's popularity grew, pressures resulted in this construction. Much of the development occurred before passage of the Georgia Shore Assistance Act; therefore, many structures are located closer to the ocean than would now be allowed.

After the act was passed, one Tybee resident tried to circumvent its intentions. The act requires a permit for certain activities between the high-water mark and the first occurrence of 20-foot live native trees. This particular resident planted a palm tree at the western base of the dunes, which allowed construction closer to the dunes than did the existing oaks. Or so the resident thought. The maneuver was detected, and no new construction was permitted.

Return to your car and continue south on Butler Avenue.

Intersection of Butler Avenue and Fourteenth Street

Turn left onto Fourteenth Street at this intersection, which has a traffic light. (Notice the drop in elevation from the intersection to the beach.) Drive into the parking area at the end of the road (a parking fee is charged during summer months) or park at one of the meters. The parking area extends to the south for four blocks from this point.

You'll see the condition of the sand dune system north of Fourteenth Street. Do the dunes begin to drop in elevation and taper toward the parking area? The dune restoration of 1989 placed sand fencing and sea oats along the entire length of the parking area. However, attempts to prompt dune development depend on the existence of a protective beach seaward of the restoration. If past erosion patterns continue, the creation of a well-developed dune system here is improbable.

Continue south along the shoreline either by walking or by driving through the parking area or down Butler Avenue to Sixteenth Street. (The Marine Science Center is located between Fifteenth and Sixteenth Streets on the beachfront.)

17.5 miles: Parking Area at Beach and Sixteenth Street

Two years after the 1976 renourishment 62

percent of the beach fill had eroded from the southern 2,100 feet of the project north of Eighteenth Street. The high-water line retreated again to the base of the 10-foot seawall, which had become completely exposed in places. (Renourishment of these beaches should occur every three years according to project plans.) The 1987 renourishment created another wide beach—at least for a time (fig. 5.14).

Loss of this portion of Tybee's recreational beach directly affects beach users, Tybee merchants, and the town's economy. The hotels, amusements, and other facilities of Tybee's business district are located here because the railroad (1887–1933) and the highway ended nearby. The large parking areas anchor this area as the island's major recreational beach (fig. 1.16).

Continue either by foot or automobile along the beach to the end of the parking area.

You can observe the rise in elevation along Sixteenth and Seventeenth Streets toward Butler Avenue. Storm waters have flooded these streets during recent northeasters. Notice, too, that the condominiums to the west are elevated to conform with requirements of the FEMA (Federal Emergency Management Agency) Flood Insurance Program. As noted in chapter 7, elevating a structure in high-hazard zones allows storm-surge waters to pass under the building and around support pilings.

The two-story brick building in the middle of the parking lot is the Tybee Island Marine Science Center. When first built, it served as the Tybee Police Department Public Safety Building. Notice its method of construction. Do you see pilings to elevate the structure? The building, originally a local government facility (and, ironically, built with the help of federal funds), is *not* in compliance with FEMA regulations! When FEMA discovered this violation of flood protection guidelines, Tybee was almost dropped from the National Flood Insurance Program. Close to 200 other local violations contributed to FEMA's stance; the police building was the most obvious infraction. Only through after-the-fact corrections of many violations and some political compromise was Tybee able to stay in the program.

17.7 miles: Intersection of Eighteenth Street and the Beach Parking Area

During periods of severe erosion, the seawall along this section may be completely exposed over its 8- to 10-foot height. Extremely high tides may top it. At least three storms during the 1980s carried wave-driven waters across the parking areas (fig. 5.14) and down the access streets. In 1983 granite boulders were put in place to protect and bolster the aging wall (fig. 5.16a).

Rapid erosion of the south end beach in the late 1970s exposed dilapidated wooden groins that had been covered by the 1976 artificial beach (fig. 5.13). The barnacle-encrusted structures became hazards to swimmers; some bathers were injured when carried by waves into the groins. Eventually, the town erected red warning flags along the groins, and the area had to be closed to swimming (fig. 5.15a). As elsewhere along Tybee's beaches, former erosion control structures became liabilities to the town; the hazards of the decaying structures caused closing of the very beaches that were intended to be "saved." Since then, most of the groins have been removed.

From the parking area's southern end, gaze over the beach to the southeast. The rubble-mound granite groin constructed during the 1986–87 replenishment is nearby, marking the restoration project's southern boundary. Like the north end groin, this new structure is designed to retain sand on the beach for a longer period of time. However, sand continues to move out of the project area.

The area south of the groin, which experiences severe erosion and is often exposed to wave attack, would benefit from any sands that might move to its shores. Residents along the shoreline to the south of the project boundary (the groin) are angry that the project did not extend all the way to the island's south tip. A brief walk along that beach shows why they're so upset.

The walk to the island's south end is one-half mile. To drive there, follow Eighteenth Street to Butler Avenue. Turn left on Butler. Follow the road as it curves to the right. At the next curve to the right, find a parking space.

17.9 miles: Intersection of Butler Avenue and Chatham Avenue at Observation Tower
Walk to the beach along the sandy trail to the south of the road.

A former observation platform serves as a landmark. Notice the evidence of erosion on the beach and behind the seawall. Compare the elevation of this beach south of the groin with that on its north side. Cycles of erosion and accretion along Tybee's south end have been attributed to changing channel positions in the ebb tidal shoals at the mouth of Tybee Creek (fig. 3.15). A 1978 study by the Corps of Engineers indicated that incoming tidal currents were responsible for sediment loss on the south end. The cyclical migration of the channels toward and away from the shoreline is thought to cause erosion and accretion. The south end groin is intended to negate the action of this nearshore channel along the beach north of the groin.

These south end shoals, or sandflats, are interesting. If you walk out at low tide, be aware of tidal stage. Make your excursion brief! Waters of the incoming tide flood the shoals quickly with strong currents!

The hammocks and barrier beaches of Little Tybee Island, a popular recreation area, are visible across Tybee Creek. The state is purchasing Little Tybee Island for conservation and recreation.

Along the river beach many individuals have constructed summer cottages as well as more permanent residences. Although "No Trespassing" and "Private Beach—Keep Off" signs may be posted along the shoreline, the area below the mean high water line is public property managed by the state. Enjoy the walk.

If you wish to see some of the marshfront areas, continue on Chatham Avenue until it ends. (You may wish to detour at either the public dock or the public boat ramp on Chatham for views of the Tybee Creek area.) Turn right at the end of Chatham on Venetian Drive.

Notice how low the upland is along the marsh edge. This area may be flooded during storm tides.

When Venetian Drive ends, turn right on Twelfth Street. Follow Twelfth Street to Butler Avenue. At Butler Avenue, turn left to depart the island via Highway 80.

Index

Altamaha River, 13, 18, 88, 89
  inlet, 89
Altamaha Sound, 88, 89
Army Corps of Engineers (U.S.), 27, 46, 49, 72, 74, 75, 78, 98, 111, 118, 119, 150, 167

beaches. *See also* control efforts; islands
  access, 1, 5, 8, 112
  accretion, 23, 25–26, 78–79, 81, 83, 92–93, 94
  development, 3, 7, 10, 38–39, 41, 54
  dunes, 20, 24, 27–28, 29, 43, 54, 61, 63, 64, 79, 82, 86, 114, 115, 150
  dynamic equilibrium, 19, 28–29, 41, 48, 54, 164
  erosion, 4, 8–9, 23–24, 26, 28–29, 55, 60, 61, 62, 65, 75, 77, 78, 82, 83, 87, 146–147
  replenishment, 4, 9, 47, 50, 72, 73, 76, 78, 95, 98–99, 115
Beach Hammock, 81
Blackbeard Island, 10, 12, 21, 26, 64, 86–87
breakwaters, 47. *See also* control efforts
building. *See* construction
bulkheads, 42, 79, 98–99, 113, 115. *See also* control efforts

Cabretta inlet, 60, 86, 87, 95
Cabretta Island, 10, 11, 21, 87, 88
Christmas Creek Inlet, 107, 110
Coastal Barrier Improvement Act (U.S.), 117–118
Coastal Barrier Resources Act (U.S.), 52, 81, 90, 92, 117–118
Coastal Georgia Regional Development Center (CGRDC), 35, 59, 118–119, 147

Coastal Marshlands Protection Act (Georgia), 81, 111, 118
construction, 117
  coastal, 121–140, 175–176
  design, 122, 125–128, 133–135
  evaluation, 129–135
  high-rise, 137–139
  mobile, 136–137
  modular, 139–140
  strengthening, 135–136
  risks, 121–125
continental shelf, 7, 13, 43–49. *See also* sea level
control efforts
  financing, 40, 51, 52, 55, 77, 117
  "New Jerseyization," 4, 5, 40, 43, 55, 76
Corps of Engineers. *See* Army Corps of Engineers (U.S.)
Cumberland Island, 1, 7, 10, 12, 41, 47, 107, 110
currents, 20–21, 26, 43, 46–47
  longshore, 20, 21, 26, 44
  tidal, 22–23

deltas/shoals, 18, 22, 23, 80, 88, 114
Dora, Hurricane, 34, 67, 68, 76, 98, 105, 106, 145
dredging, 20, 23, 39, 48, 50, 61, 72, 75, 77, 79, 101, 111, 112, 118
dunes. *See* beaches
dynamic equilibrium, 19, 24, 27–28, 48, 61

East Beach, 96, 98, 99, 100
ecology, 163
equilibrium. *See* dynamic equilibrium
erosion, coastal, 1, 4, 23, 24, 41–47, 49–50, 54, 61, 66, 113. *See also* beaches; individual island names
  and control efforts, 4, 8, 9, 10, 39, 40–55, 61, 65–68, 76, 93, 94–95, 98–100, 112, 113, 149, 167–168
  risk factors, 69
  and sea level rise, 23
  signs, 24, 60, 62
  storm related, 23, 58
estuaries/embayments, 13, 17, 114

Federal Emergency Management Agency (FEMA), 58, 59, 116, 149, 151
finger canals, 63
flooding, 4, 30, 58, 59, 62, 74, 79, 80, 144–145. *See also* National Flood Insurance Program
Fort Screven, 9, 66–67, 76, 80
Frederica River, 96

geologic history. *See* history
Georgia Bight, 20, 22
Goulds Inlet, 26, 94, 95, 96, 97, 98, 99, 100, 107
groins, 8, 23, 42, 45–47, 59, 73, 75, 76, 77, 78, 95, 113, 114. *See also* control efforts
groundwater, 59, 60
Growth Management Strategies Act (Georgia), 118
Guale Coast, 7, 13

Hampton River, 88, 89, 92, 93, 95
  inlet, 89, 93, 95
history, 76, 89
  geologic, xiii, 7, 13, 14, 16, 17, 18, 58, 70, 80, 83, 84, 87
  human, 7–8, 10–12, 71, 81

Hog Hammock, 7, 88
Hugo, Hurricane, 28, 30, 35, 36
hurricane(s), 1, 34, 79, 100–101, 144–145. *See also* individual hurricanes
  barometric pressure, 124–125
  checklist, 141–143
  of 1893, 28, 72, 79, 80
  of 1898, 28, 32, 33, 79, 87, 89, 105, 145
  evacuation, 4, 35, 36, 79–80
  formation, 31, 33
  prediction/probability, 4, 31, 32–35, 59
  Saffir-Simpson Scale, 31–32
  storm surge, 31–32, 58, 79, 87, 100–101, 123–124
  track, 31, 33, 35, 144, 166

Indians. *See* Native Americans
inlets, 17, 21, 22, 52, 60–61, 79, 81, 89. *See also* individual inlet names
  migration, 78, 87, 89, 94, 95, 96, 98, 100, 107
insurance. *See* National Flood Insurance Program
islands. *See also* beaches; individual island names
  access, 1, 5, 8, 58, 81, 86, 88, 89, 156
  barrier, xiii, 1, 5, 6, 13, 17, 161
  development, 3, 5–6, 8, 52–53, 58, 65
  history, 7, 159–160
  migration, 17, 18, 24, 27, 78, 85, 95, 148
  overwash, 29, 30, 79, 89, 95, 100
  pollution, 4, 63, 118
  undeveloped, 10, 12, 118

Jekyll Island, 1, 3, 7, 8, 9, 19, 23, 61
  access, 1, 5, 8, 101

development, 3, 8, 10, 37, 102
erosion, 10, 11, 44, 55, 68, 101, 104, 105
evacuation, 106
flooding, 4, 9, 48, 105
hazard-risk evaluation, 103
management, 4–5, 7, 8, 10, 106
jetties, 47, 110, 113, 114, 115

land use, 3, 4–6, 8, 9, 12, 39–44, 52, 57, 64–65, 80, 89, 106–107, 111–119, 155, 168, 173
Little Cumberland Island, 10, 12, 41, 107–109
Little St. Simons Island, 10, 12, 88, 89, 90, 91
Little Tybee Island, 10, 11, 41, 70, 72, 80, 81, 144

MacKay River, 96
marsh. *See* salt marsh
mobile homes. *See* construction

Nanny Goat Beach, 4, 61, 86, 87
National Flood Insurance Program, 81, 116–118, 125
National Weather Service, 31, 35
Native Americans, 7, 10, 81, 89, 90
northeasters, 20, 28–30, 79
  Ash Wednesday Storm, 29

Ogeechee River, 13
Ossabaw Island, 7, 10, 11, 83
Ossabaw Sound, 83
overwash. *See* islands

parks and recreation areas, 8, 9, 11, 71, 77, 81, 98, 101, 112, 154, 160

permits, 111–115, 118, 147, 150, 156

regulations, 101, 111–119
  access, 112
  construction, 114
  dredging, 111, 112, 118
  dune, 114–115
  insurance, 116–117
replenishment. *See* beaches
revetments/riprap, 10, 25, 44, 45, 76, 113, 115. *See also* control efforts
rivers, 13, 15, 19–20, 62, 63, 72, 78, 79, 83. *See also* individual river names

St. Andrew Sound, 107
St. Catherines Island, 7, 10, 12, 23, 84–86
St. Simons Island, 1, 3, 4, 5, 7, 8, 9, 10, 90, 96–101
  access, 1, 8, 9, 96, 101
  accretion, 92
  development, 3, 8, 9, 37, 96, 97, 101
  East Beach, 96, 98, 99, 100
  erosion, 10, 44, 45, 55, 61, 67, 92, 96–97, 101, 105
  evacuation, 101
  flooding, 100, 105
  hazard-risk evaluation, 97
  management, 9, 101
  Massengale Park, 96, 97
St. Simons Sound, 100, 104
salt marsh, xiii, 13, 18, 22, 37, 64, 70–71, 80, 83, 88, 96, 111–113
sand supply, 19–20, 23, 48–50
Sapelo Island, 4, 7, 8, 10, 11, 12, 21, 61, 87–88, 145
Savannah River, 13, 20, 61, 70, 72, 77, 78–79

inlet, 71, 72, 78–79
Sea Island, 1, 3, 4, 5, 9, 90–96
  access, 1, 5, 8, 92, 95
  accretion, 21, 25, 26, 93
  development, 3, 8, 37, 92
  erosion, 9, 25, 29, 43, 93, 95
  evacuation, 95
  hazard-risk evaluation, 91
  management, 5, 9, 95–96
sea level, xiii, 1, 7, 14–19
  fluctuations, xiii, 7, 13–14, 15, 16, 18, 162
  rise, xiii, 1, 4–7, 9, 16, 18–19, 23, 60, 162
seawalls, 4, 8, 9, 10, 11, 25, 29, 42, 43, 73, 75, 76, 93, 114. *See also* control efforts
sediment transport. *See* currents
Shore Assistance Act (Georgia), 114–116, 118, 150
site selection, 39, 57–68, 110, 121, 172
  rating system, 69
soil, 63, 68, 157
spits. *See* beaches: accretion
storms, 1, 4, 27, 33, 77, 79, 100, 123, 124, 144, 158. *See also* hurricane(s); northeasters
structural engineering. *See* construction

tides, 21–23, 29, 31, 144
Tybee Creek, 70, 72, 74, 76, 79, 81
  inlet, 70, 72, 74–75, 76, 77, 81
Tybee Island, 1, 3, 4, 5, 8–9, 42, 61, 63, 70–80
  access, 1, 3, 58, 65, 71, 79
  accretion, 23, 72–73, 76
  development, 3, 8, 9, 37, 58, 71
  erosion, 4, 20, 23, 41, 46, 48, 51, 55, 70, 71, 72, 73, 74–76, 78, 79
  field trip, 179–186
  flooding, 29, 72, 79
  hazard-risk evaluation, 70
  management, 5, 80

vegetation, 1, 24, 36, 38, 61–64, 82, 83, 86, 89, 101, 113, 114, 158, 164

Wassaw Sound, 81
waves, 20, 22, 27, 43, 75, 116, 124, 145. *See also* erosion
wetlands, freshwater, 37, 86, 101, 119. *See also* salt marsh
wildlife, 6, 8, 11–12, 81, 83, 84–85, 87, 158
Williamson Island, 10, 11, 30, 70, 80–81
wind, 20, 28, 137–138. *See also* hurricanes; northeasters; storms
Wolf Island, 10, 12, 88–89

Library of Congress Cataloging-in-Publication Data
Living with the Georgia Shore / by Tonya D. Clayton . . . [et al.].
p. cm. — (Living with the shore)
"Sponsored by the National Audubon Society."
Includes bibliographical references and index.
ISBN 0-8223-1215-8 (alk. paper). — ISBN 0-8223-1219-0 (pbk. :
alk paper)
1. Coastal zone management—Georgia. 2. Barrier islands—
Environmental aspects—Georgia. I. Clayton, Tonya D.
II. National Audubon Society. III. Series.
HT393.G4L58 1992
333.91'7'09758—dc20   91-41261 CIP